LAND IS LIFE

DISCARDED

HD
1332
L36
1992

LAND IS LIFE

Land reform and sustainable agriculture

Edited by NIGEL DUDLEY, JOHN MADELEY
and SUE STOLTON

Intermediate Technology Publications
in association with Foundation Development and Peace 1992

Intermediate Technology Publications Ltd
103–105 Southampton Row, London WC1B 4HH, UK
Published in Germany by Foundation Development and Peace
(SEF)

© Stiftung Entwicklung und Frieden (SEF), Bonn 1992

A CIP catalogue record for this book is available from the British
Library.

ISBN 1–85339–146–8

*The publishing of this book was made possible by financial support of
the Ministry of Environment, Lower Saxony and the Senator for
Economics and Technology, Berlin.*

*The publishers would also like to acknowledge financial assistance
from the Right Livelihood Foundation.*

Typeset by J&L Composition Ltd, Filey, North Yorkshire
Printed in Great Britain by BPCC Wheatons, Exeter

Contents

Notes on contributors

MARTIN ADAMS is a rural development planner

PETER BUNYARD is an editor of *The Ecologist*

HELGE CHRISTIE co-ordinates the GATT campaign in Norway

JOHN CUSWORTH works for the Development and Project Planning Centre at the University of Bradford

LINO DE DAVID works for the Centre of Popular Alternative Technology, Brazil

NIGEL DUDLEY works for the London-based Earth Resources Research

MARK FEEDMAN works for the US-based Rural Development Service Group

LARRY LOHMANN is an associate editor of *The Ecologist*

JOHN MADELEY is editor of *International Agricultural Development*

DR UWE OTZEN is an agricultural economist with the German Development Institute

DR VITHAL RAJAN works for WWF International

SUE STOLTON works for the Equilibrium research organization

DR MOHAMED SULIMAN is director of the London-based Institute for African Alternatives

ROMY TIONGCO is a lecturer at the University of Edinburgh

LAWRENCE WOODWARD is director of the Elm Farm Research Centre in England

Foreword

'For us the land is our life; we would defend it to the last drop of our blood' Filipino peasant

When land in the Piaui district of north-east Brazil was redistributed to small farmers, the expected happened. Ownership of the land spurred a new interest among the farmers. On rainfed farms, yields increased by between 10 and 40 per cent; on irrigated fields, they rose by between 30 to 70 per cent. 'Clear title had given the farmers an incentive to invest in the land and adopt better technologies'.[1]

In Thailand, people who were encroaching on forest land have been given security of tenure over that land; they are now planting trees to replace the old ones that have been axed. In Zimbabwe, a survey found that the transfer of land from large-scale farming to smallholders increased productivity.

From many parts of the world are coming accounts of the close link between land reform and sustainable agriculture and forestry. The reaction of farmers in Piaui to land reform is by no means exceptional. If it could be replicated throughout the developing world, then a substantial and sustainable increase in food output is possible from existing lands. It may be too early to say if the higher yields that followed land reform in Piaui will be sustained. But ownership of land clearly gives farmers the incentive to farm it in a way that should allow yields to be sustained.

The world now faces a desperate situation over food availability for millions of people, many of whom are hungry to the point of starvation. And with little 'new' land available to bring into production, world population is set to increase by 50 per cent over the next 30 years. Many people starve because of lack of ability to buy food, rather than from the inadequate supply. But while redistribution of income is vital for the poor, increases in food output are also vital if starvation is not to engulf ever-growing numbers of the world's poor. Of equal importance is that increases are achieved in a way that can be sustained. There is an urgent need to consider how the ownership of land affects food production, and for land to be owned by people who are likely to get the most out of it in a sustainable way.

Many millions of people in the developing world are either landless or work on land that is owned by others. The effects of landlessness on both the landless and their environment can be devastating. Landless people often rank among the poorest of the poor; those who live near tropical forests, and who have no land, are more likely to move in after the loggers and clear a small plot to grow crops; the forest suffers and the land may only support crops for a short time. Farmers who rent land and who have no security may not be interested in working it in a sustainable way; they want to get all they can from the soil in the short run. So land could well be pumped with chemicals, and essential maintenance neglected; the ability of that land to produce for future generations could be eroded.

On the other hand there is little point in giving people access to land if they have neither the skills nor the money to farm it adequately. Some well-meaning land-reform programmes have failed because the people who were given back land were unable to make a satisfactory living. There are two main reasons for this. First, many displaced communities have lost their traditional skills in farming and gardening. Just as important, the modern, intensive agricultural methods introduced into many countries rely on expensive chemicals, energy and machinery, putting them out of reach of the poorest peasants. Having access to land without the capability of using it properly is little better than having no land at all.

Various methods of organic, ecological or biological agriculture offer a way out of this dilemma. Simple to practise, cheap in terms of external inputs, and using systems that do not damage the environment, these methods offer real hope to the poor and dispossessed. When, however, outsiders own land the situation may be perceived by those who farm it as unjust, and it may not be worked in a sustainable way. 'Genuine land reforms have led to greater agricultural production because they have redressed the inefficiencies of inequality', point out Frances Moore Lappe and Joseph Collins.[2]

Jigsaw

It is now widely recognized that it is little use increasing food output in a way that cannot be sustained. People who own, or otherwise have secure tenure to the land they farm usually care deeply about it and are more likely to farm it sustainably. They want their land to provide for themselves today and for their children tomorrow. When land reforms happen, food output increases and is likely to increase in a sustainable manner. Land reform therefore looks more and more like a vital piece of the sustainable agriculture jigsaw. But although land reform has been much praised for its potential, it is usually ignored by governments because it might require difficult political decisions.

This book explores some of the issues involved. It draws on ideas from a conference held in Berlin in November 1991, 'Soil for Life: Promoting Sustainable Land Use', which was organized by two non-governmental organizations (NGOs), the German-based Stiftung Entwicklung und Frieden (Foundation Development and Peace) and the Swedish-based Right Livelihood Award, in co-operation with the Senator for Economy and Technology, Berlin, and the Minister for the Environment, Lower Saxony. Their aim in holding the conference, probably the first international conference to link up the issues of sustainable agriculture and land reform, was to highlight concern that these issues are not featuring prominently enough on the international agenda, and, in particular, could be neglected by the United Nations Conference on Environment and Development (UNCED), held in Brazil in June 1992.

Although there have been few studies of how land reform might contribute to sustainable agriculture, there is enough evidence to show a direct link between the two. In his overview of the issues, Nigel Dudley points out that most good agricultural land is owned or controlled by a small minority of rich landowners, and that despite years of trying to achieve land reform, this has not occurred in significant ways in most Third World countries. The chances of implementing policies that are capable of feeding growing populations, 'and maintaining a sustainable agricultural system, seem increasingly to rely on some form of redistribution of land', he says. But Dudley warns that land reform will only be possible if the poor and the powerless 'have access to production methods that do not rely on machinery and agrochemicals they cannot afford'.

Nature has no room for manoeuvre, points out Dr Uwe Otzen, an agricultural economist with the German Development

Institute. Millions of tons of soil are being lost each year, he said, because of soil erosion: 'if land continued to be degraded then little will grow'.

Dr Mohamed Suliman, of the London-based Institute for African Alternatives, says that elevating sustainability to a universal goal is perhaps the greatest achievement of the Brundtland Report, *Our Common Future.* The cumulative effect of land misuse is a major reason for the rampant poverty of rural Africans, says Suliman. Writing of the land-use conflict in Sudan between cotton and wheat, and about the growth of sorghum for export, he opens up the whole controversy over whether exporting crops means that farmers are more likely to farm in a way that damages the soil. In Africa it has been observed that mistreatment of the land tends to increase in proportion to the farmer's commitment to cash cropping.[3] When too much land is under cash crops, sustainability is at risk.

Political change

Dealing specifically with land reform aspects, Ivo Poletto and Lino de David write with certainty that when land reform takes place the farmers will produce more food and will do so in a way that can be sustained. They point out that 70 per cent of rural households in Brazil own no land. Without agrarian reform, they say, the Amazon rainforest cannot be preserved, as people with no land would simply continue moving into forest areas. And they stress the need for 'deep-rooted political change . . . if agrarian reform is to be achieved. Land reform cannot happen unless we have the political prerequisites for it'.

In Colombia the government has come under intense pressure from agricultural, mining and logging interests to open up more of its forests for exploitation. Peter Bunyard tells of how, instead of giving in to the pressure, the government embarked on an imaginative land reform programme in the shape of granting titles to the Indians of the Amazon; the reform has embraced some 18 million hectares of land. The Indians farm their forests intensively but sustainably and the granting of land title will mean that their forest habitats will be preserved for them to continue their practices.

As noted above, access to land is pointless without the ability to use it successfully. In the Dominican Republic,

Mark Feedman tackled the problem from the other way round; he started a local teaching initiative in the highlands, introducing villagers to the concepts of biodynamic agriculture. His presentation shows how the grassroots approach is having ramifications beyond the country's borders.

Writing of India, Vithal Rajan says that although the power of landlords in rural India is strong, village women are empowering themselves in some areas through the formation of their own organizations. By harnessing their power, they are getting better access to land. Not only are they organizing themselves into efficient units to practise banking and education within their villages, they are also repossessing derelict land and introducing new sustainable forms of agriculture. Significantly, the 'new' methods of biological and cultural pest control being introduced by Indian research organizations are almost identical to the traditional methods, which, just barely, remain within the memories of older villagers.

Larry Lohmann illustrates very clearly the inter-relationships in Thailand between a secure title to land and the confidence to practise agricultural systems. The connection between sustainable agriculture and the control of land is seen by the fact that resistance to expropriation of land is often strongest in places where traditional or organic forms of agriculture are long established, and therefore people have a correspondingly deep relationship with the land on which they live. This is clearly demonstrated by an example of resistance to land grabbing in an area of North Thailand where traditional irrigation systems have been used for hundreds of years.

The experience of farmers in the Philippines shows that only when they have secure title to land do they have the incentive to start practising sustainable agriculture. Romy Tiongco, who has worked in the country for many years, describes how farmers and campaigners for land reform worked together to secure land and in so doing defeated soil erosion problems in one area of the islands.

Drawing on his experience of land reform in Zimbabwe, John Cusworth highlights the complexity of the issues. He points out that a constraint on the country's resettlement programme was whether farm productivity could be sustained under smallholder agriculture. Large-scale farming makes a sizable contribution to the national economy, he says, and there are uncertainties over whether small producers could maintain this. Some studies, however, support the view that the transfer of land from large to small farmers increases productivity. Cusworth stresses that land reform needs a series of measures to go with it, such as credit, improved availability of farm inputs and better marketing.

Namibia has some of the largest landholdings in Africa. Martin Adams writes of how when the country came to independence in 1990, the South West Africa People's Organization announced its intention to 'transfer some of the land from those with too much of it to the landless majority'. But land reform has proved difficult in a ranching country where there are strong interests in maintaining the status quo; Adams examines some of the complexities involved.

The involvement of Western countries in these issues is explored from different viewpoints. As organic agriculture should not be confined to the South, Lawrence Woodward describes the current situation in Western Europe, raising questions about the maintenance of good organic farming systems. Helge Christie looks beyond the farm gate and considers what people in European countries are doing to lobby for more equitable trading conditions. And Nigel Dudley summarizes conditions regarding organic agriculture in the newly democratic countries of central and eastern Europe.

Most of the articles in this book were presented at the 1991 Berlin conference but it also includes a number of contributions from people who were unable to be present. The conference ended by adopting a statement which said that the present world agricultural system is 'increasingly unstable. It is degrading the resource base and poisoning the environment. . . . Moreover the promises of the present system have not materialised for the majority of the world's people. All the world's people can be fed if the North and the South adopt sustainable agricultural policies.' These could best be achieved, says the statement, when people have equitable rights to land and other agricultural resources.

For many millions of people, the land they farm is their life — it is the resource that provides them with the means of life. But land can only continue to be life if it is farmed in a sustainable way. More work needs to be done to explore the links between land reform and sustainable agriculture and to document specific examples. It is our hope that this account will help to stimulate an ongoing debate on the issues. With awareness growing, both of the importance of land reform and of the need to grow food in a way that can be sustained, these issues will be at the forefront of the food and agriculture debate in the 1990s and beyond.

NIGEL DUDLEY, JOHN MADELEY and SUE STOLTON
January 1992

PART 1: *Stabilization of Agricultural Resources*

World Hunger, Land Reform and Organic Agriculture*

NIGEL DUDLEY

The thesis I wish to present here is that there needs to be a greater, or at least a more conscious, link between two agricultural reform movements working in the South: the organic or biological farming movement, and the many groups and individuals working towards land reform, particularly with respect to farmland. Most analyses of future prospects for agriculture in the less developed countries are either despairingly pessimistic or base their limited optimism almost entirely on technical approaches to current or projected short-falls in food supply, through the 'modernization' of agriculture.

Both approaches have their limitations. Pessimism is understandable, given the mounting problems of malnutrition and environmental decay, but doesn't do much to help people facing the daily problems caused by shortages of food, fuel and other necessities. Technical fix enthusiasts tend to ignore the political issues which help cause or increase food shortages, so that their suggested remedies either fail to take these into account or, in some cases, make them even worse. Most technical, or rather high-tech, solutions also sidestep many of the environmental side-effects of the agricultural practices they are promoting. This chapter starts from the perspective that other production methods will only be effective in feeding a population if they also take environmental and political considerations into account.

---

* Research assistance from Sue Stolton

1

Population, politics and food shortages

The famine equation

For the past twenty years, it has been argued that a combination of rapid population growth and the lack of sophisticated agriculture in the South is causing a steadily escalating spiral of food shortages, malnutrition and famine. For simplicity, we will call this argument the famine equation:

overpopulation + poor agriculture = famine

The relative importance given to the two components alters, depending on who is arguing the case. In the 1970s, the population issue was highlighted by the fledgling environmental groups and through the writings of biologists such as Paul Erlich of Stanford University, whose polemical tract *The Population Bomb* (Erlich 1968) was a bestseller. In this deliberately populist book, Erlich argued that if current rates of population increase continued, there would not be enough existing or potential farmland to provide sufficient food for people, leading to widespread famine, political unrest and social breakdown.

In the same year that Erlich's book was published, a group of Italian industrialists belonging to the informal Club of Rome commissioned a team at the Massachusetts Institute of Technology to prepare a report on the effects and limitations of continued worldwide growth. This included an examination of the relationship between population and food supply, and further focused attention on the issue:

> If the present growth trends in the world population, industrialization, pollution, food production and resource depletion continue unchanged, the limits to growth on this planet will be reached sometime within the next hundred years. The most probable result will be a rather sudden and uncontrollable decline in both population and industrial capacity.
>
> (Meadows *et al*. 1972)

During this period, a new breed of 'development experts' was also arguing that food shortages in the South were being exacerbated by unsophisticated and inefficient farming methods. The main problems were identified as low-yielding crop varieties, lack of efficient fertilizers, poor irrigation, heavy crop losses through pest and disease attack, and a lack of understanding of basic agronomics.

This thesis gained support from international bodies such as the United Nations and was developed in a number of influential reports; for example the Independent Commission on International Development Issues, chaired by Willy Brandt, recommended that: '. . . special attention should be given to irrigation, agricultural research, storage and *increased use of fertilizer and other inputs* . . . [our emphasis]' (Brandt 1980).

Limitations to the famine equation

The twin responses to the problem of insufficient food supply were therefore widely seen as 'controlling' population growth and improving the productivity of agriculture. Programmes aimed at achieving this became a major concern of the Northern governments and international aid agencies.

However, this analysis did not go unchallenged, and attracted criticism from a growing number of development specialists. They argued that although both population growth and poor agricultural practice are important factors in the problem, uncritical acceptance that they were the entire problem led to the two following incorrect and damaging assumptions:

o *the victim is to blame*, that is, hungry people were seen as being the cause of their own problems through having too many children; and
o *salvation via the magic bullet of technology*, that is, the only way out of the mess caused by overpopulation was seen as rapid intensification of agriculture.

Despite an analysis that appears obviously flawed in retrospect, the famine equation was seized on with enthusiasm by governments and industry in the North. This is hardly surprising. The suggested solutions offered the booming post-war industry of Europe and North America, and especially the powerful agribusiness corporations, virtually unlimited new markets in the former colonies, with the added cachet that a profitable business might also help feed hungry people.

The famine equation remains a powerful concept today. Public statements from pesticide companies, for example, often argue that agrochemicals are a vital tool in alleviating world hunger. Uncritical acceptance of the basic premise has coloured virtually all responses to problems of hunger and the Third World.

3

The Green Revolution

The first major result of this thinking was the so-called Green Revolution, which was an attempt to meet future food requirements by improving the efficiency and output of agriculture; of filling 'millions of rice bowls once only half full', according to one expert (Jackson 1990). The Green Revolution was followed through most intensively in Asia, although it has had ramifications for almost all Third World countries. Three main aims can be identified (Conway and Barbier 1990):

○ the production of early maturing, day-length insensitive and high yielding varieties (HYVs) of staple foods such as rice;
○ the distribution and promotion of inputs to further increase productivity, such as soluble fertilizers, and pest and disease-controlling pesticides;
○ the implementation of strategies to promote use of HYVs and agrochemicals in the most favourable geographical areas and amongst those best able to benefit.

The Green Revolution was promoted and to some extent implemented by the Western nations, and aimed especially at the Asian countries. It was founded on an uneasy mixture of altruism and business interests, with genuine attempts to improve food production becoming confused with corporate imperatives. A representative of the Rockefeller Foundation, which sponsored much of the research and promotion of the Green Revolution, outlined his philosophy to the US Foreign Affairs Committee in 1969:

> The expansion of world economies through the intensification of agriculture may also provide major and direct benefits to the developed nations, particularly in the expansion of cash markets for an ever broader range of industrial products.
>
> (Harrar 1969)

The failure of the Green Revolution to live up to its early claims is the key to much of what follows.

Seven problems with the Green Revolution

On one level, the introduction of HYVs has been a spectacular success. Between one-third and one-half of all rice areas in the developing world are planted with HYVs. Per capita rice production in Asia has increased by 27 per cent (Conway and Barbier 1990). The Green Revolution has introduced

4

profound changes to the practice of agriculture throughout most of the South.

However, it is now almost universally accepted that these changes have not necessarily done much to improve the average diet of the people who live there. The initial euphoria about the potential of the Green Revolution began to be questioned in earnest during the 1970s (e.g. George 1976), although the momentum of governments and aid agencies meant that developments continued as usual for some time afterwards.

Although the debate about the effectiveness of the Green Revolution is now well known to most people working on development issues, it is nonetheless so central to the arguments presented below that a brief restatement is necessary. This is particularly true because the problems created by the introduction of HYVs have had impacts on both land ownership and sustainability of agriculture, the two key issues being addressed here.

New crop varieties need large inputs of agrochemicals
The HYVs were bred to respond quickly to high levels of artificial, soluble fertilizers, which they require to be fully productive. In addition, the emphasis on yield in the breeding strategy has meant that some of the pest and disease resistance found in traditional varieties has often been lost. The pest problem has been increased by the practice of planting the same few varieties over huge areas.

This means that use of agrochemicals increases sharply when HYV strains are introduced. Initial use of HYV rice in the Philippines boosted production spectacularly, but headway was lost again when the rice was found to be more susceptible to disease (Brown and Eckholm 1974). Use of fertilizers often rose four to five times with the new varieties (George 1976).

Agrochemicals are expensive   Even if we accept that agrochemicals are on balance beneficial, they are expensive. Experience has shown that in many cases only the richer farmers are able to afford them, and thus afford to grow the HYVs, even if the latter are subsidized by governments or industry. Richer farmers produce higher yields per unit area, while their poorer neighbours maintain the original production levels. Net food prices fall and the poorest farmers get even less for their produce than before.

Small farmers are forced out of production   One result of
the proportional increase in profits for the richest farmers is
that they are able to buy more land. This is usually from
poorer farmers who have lost out in the Green Revolution,
and can therefore no longer compete satisfactorily. A side-
effect of the introduction of HYVs has been the concentration
of good agricultural land in fewer hands. This result was
recognized right at the beginning of efforts to launch the
Green Revolution, as shown by a further quotation from the
US Foreign Affairs Committee referred to earlier, this one
from a member of the Inter-American Development Bank:

> The tenants, I think, may become a diminishing breed as they get
> squeezed out gradually by landlords reclaiming their holdings
> because agriculture has become profitable. . . . The GR might
> make the rich still richer and enable them to capture markets
> previously served by smaller semi-subsistence farmers.
>
> (Carroll 1969)

Richer farmers concentrate more effort on cash crops
The change from smaller to larger farms has had an impact
on the type of food grown. Larger farmers usually put less
effort into producing crops for local distribution. Over the
past few decades, the greatest profits have come from exports
of cash crops such as coffee, tea, soya (for animal feed), cocoa,
beef and tobacco. Increased productivity through HYVs does
not necessarily mean increased production on a national level.
As use of HYVs allows greater concentration of landownership, this can also lead to a widespread switch to cash cropping
for export or for internal sale. Either way the poorest peasants
and landless workers lose out in terms of food availability.

The cash cropping argument is illustrated by food production in Kenya between 1976 and 1982. Over that period
marketed coffee increased by 10 per cent, sisal by 49 per cent,
tea and cotton both by 54 per cent and sugar by 88 per cent.
Marketed maize, on the other hand, stayed roughly the same,
and the amount of rice sold actually declined (UNICEF 1984).

There is an increased emphasis on monocropping
Smaller subsistence farmers tend to grow a range of foods,
for consumption by the farmer or for sale locally. Even
varieties of staple crops such as rice are likely to vary between
different regions and, to some extent, be adapted to local
conditions. Another result of the Green Revolution has been

6

the increase in monocropping, both because farmers switched to a few HYV varieties that gave high yields and because there has been an increase in cash crops following the political impact of the Green Revolution. It is now well recognized that monocropping is both less environmentally stable (Gordon and Barbier 1990) and also that large areas of single crops tend to need higher inputs of agrochemicals.

The Green Revolution only works well on the best land High-yielding varieties not only require large inputs of agrochemicals, but are also bred for use on the best land. This further disadvantages the poorer farmers, who are likely to be forced on to more marginal land, and also means that the impact of the Green Revolution has been far less in places where good agricultural land is scarce, such as some parts of Africa.

Agrochemicals cause health and environmental problems In 1982, a report from the British aid NGO Oxfam suggested that 10,000 people a year were dying as a result of pesticide poisoning, with the vast majority of deaths occurring in the developing world (Bull 1982). More recently, large-scale problems of water pollution from fertilizers have also been identified (Pretty and Conway 1991). These issues are discussed in more detail below. The increase in agrochemicals, that is essential with the use of HYVs, has exacerbated these problems.

We might well ignore some or all of these objections if the initial aim of securing increased food supplies had been achieved. However, this is not the case. Overall food production per capita in the South has only increased by 7 per cent since the 1960s (Conway and Barbier 1990), and food production per capita has actually declined in Africa. There are now also signs of diminishing returns from some HYVs. The social and environmental costs of the Green Revolution have not always been balanced by access to food amongst the poorest people.

An overview of the Green Revolution
To summarize: conventional attempts to solve the problem of food shortages by technological means have failed because they didn't look at the wider political or environmental impacts of what was being suggested. From our perspective

here, the key results have been: an increase in concentration of landownership; and a decrease in environmental stability and the sustainability of farming, through higher uses of agrochemicals, increased monocropping and the further displacement of poor farmers into marginal areas. In many cases this has meant a decrease in food supply to the poorest people.

This is not to suggest that all technological innovations have negative effects, or that there is no role for crop breeding in agricultural development. It does mean, however, that the original, very optimistic hopes for the potential of the Green Revolution have been disproved, along with the analysis of food problems, which we have called the famine equation. This has led development experts to look at alternative explanations for current food problems and, in turn, for other possible solutions.

Towards a new theory of food shortages

Famines are caused by the grain trade Bertolt Brecht

A new thesis has emerged in opposition to the famine equation. This challenges old assumptions about population and agricultural productivity and suggests some alternative, or at least additional, causes of food problems. These are outlined briefly below, but first we look more closely at one of the original components of the famine equation: population.

Population

We started with the assumption that population was an underlying cause of famine. If this is really the case, control of population must be a central facet in any attempts to tackle the problem of world hunger; on the other hand, if population is not an underlying factor, it raises a lot of questions about why people are going hungry and why they do not have sufficient quantities of land.

The overpopulation thesis still has many capable and persuasive proponents. A recent book by Paul and Anne Erlich again puts the case that overpopulation is a potentially apocalyptic problem. In a chapter entitled 'Why isn't everyone as scared as we are?' they argue that: 'In the early 1930s, when we were born, the world population was just two billion; now it is more than two and a half times as large and still growing rapidly' (Erlich and Erlich 1990). They go on to look at food supply and conclude that the famines currently being experienced in a number of African countries are the result of too many people competing for too little land.

This is a seductive argument. There is no doubt that if global population were to continue growing at the same level as it has done over the last few decades there would come a time, sooner than we may like to think, when there would be an overall shortage of space to grow food to feed people. Thus far, the overpopulation thesis is correct. Population growth in some individual areas may already be a factor in problems of food supply, and has undoubtedly had an impact on land availability in some places.

However, there is strong evidence to suggest that it is a serious oversimplification to argue that current, or medium-term future food shortages are caused solely or mainly by space limitations; problems of distribution and accessibility to food remain more important. Far from this argument being confined to a few radical social critics, as was the case in the 1970s, it has gained a number of powerful advocates, including the World Bank.

Poverty and Hunger: Issues and Options for Food Security in Developing Countries was published by the World Bank in 1986. The report argues that, at a global level, the United Nations Food and Agriculture Organisation (FAO) has estimated that the world's population grows enough grain to provide the current population with 3,600 calories a day, well above the 2,400 calories needed to maintain a nutritious diet. Despite a virtual doubling of the world population between 1950 and 1983, per capita food production increased during this period. In a Third World country, the food that the hungry lack is usually less than 5 per cent of the country's overall food supply but: 'this does not mean, however, that a 5 per cent increase in food supplies would eliminate malnutrition. It means merely that in many countries the supply of food is not the only obstacle to food security' (World Bank 1986). Indeed, it is salutary to note that some European countries, including especially Britain and the Netherlands, are amongst the most overpopulated in the world. Cropland per capita remains far greater in most developing countries than in central Europe.

More concrete reasons for hunger are rooted in people's

ability to obtain food. Amayarta Sen, an Indian economist, identifies four ways in which people can get 'entitlements' to food in *Poverty and Famines: An Essay on Entitlement and Deprivation* (Clarendon Press 1981). These are by:

o growing their own food
o working for money and buying food
o trading or bartering for food
o being given or lent food.

Famines occur, he argues, when people lose their entitlements to food, through losing land, not having enough money to buy food and having nothing to barter. Grain exports continued from Ethiopia throughout the recent famine, a far from unusual occurrence which flies in the face of Western preconceptions about food shortages.

Other factors do play a part of course, but these are increased by the inability of the poor to buy their way out of trouble. A drought in one area of the country, or one region, might not cause a problem if food could be shipped in from other areas, but reaches crisis point if distribution systems break down, or have never been properly established.

Poor transport infrastructure and corruption, for example, are also important in stopping food from getting through to crisis points. The minor famine in Tanzania in 1985 was not due to an overall lack of food in the country but was caused by the roads not being in good enough shape to shift sufficient quantities of food around fast enough. Major and minor corruption is an important cause of slowing food distribution at a time of famine, and by no means all donor food reaches the hungriest people. Population is just one factor amongst many.

Agricultural improvements
The second element in the famine equation is the need to improve the efficiency of agriculture in the Third World as a response to growing populations. Again, at first sight this can seem entirely reasonable, and there is no doubt that many Western consultants are involved in such developments for the best of reasons.

However, when we look back at the achievements of agricultural intensification in the South we find a similar story to the problems currently being experienced by chemical

10

agriculture in Europe and North America, with the added caveat that failures in a tropical climate can have faster and more devastating effects. A mixture of environmental, social and political factors is involved. Three key issues are identified:

Not all agricultural modernization has actually been an improvement for the majority of people in the South. Indeed, countries in the South are littered with areas laid waste as a result of the failed experiments of experts from the North. For example, irrigation schemes can, if not properly managed, lead to salinization of cropland and in time to further desertification.

It is estimated that some 40m ha of irrigated cropland throughout the world's dry areas are affected by salinity, alkalinity or waterlogging, and about 500,000 hectares become desertified every year (Grainger 1990). Some of the early attempts at the Green Revolution resulted in crop losses, when existing crop strains were treated with subsidized fertilizers and grew so fast that they keeled over (George 1976). Pesticide resistance is now undermining many of the attempts to increase food production through agrochemicals (Hurst *et al.* 1991).

Most farming systems promoted in the South are more suitable for large farmers than for smallholders or peasants. As the World Commission on Environment and Development (WCED) stated in 1987:

> Agricultural support systems seldom take into account the special circumstances of subsistence farmers and herders. Subsistence farmers cannot afford the high cash outlay of modern inputs. Many are shifting cultivators who do not have a clear title to the land they use. They may plant a variety of crops on one plot to meet their own needs, and are thus unable to use methods developed for large stands of a single crop.

Forcing farmers into growing a few, high-yielding crops is dangerous. One result of the Green Revolution, and of contemporary agricultural extension work, has been to reduce the diversity of crops grown. Whereas some areas of the Third World have little tradition of settled agriculture, others already have successful farming practices, sometimes stretching back for thousands of years (Myers 1984). These have tended to disappear under paddy fields of HYV rice and

11

hectares of wheat, and monocrops of other staple foods such as cassava and maize.

Output from monocropping has been found to fluctuate more than is the case with traditional crops, because yields are more susceptible to changes in inputs. These may, in turn, be caused by variations in short-term availability of money amongst farmers; if they are unable to buy the relatively high-priced inputs the effects will be greater than if they were growing traditional varieties (Conway and Barbier 1990).

Monocropping also increases the risk of serious or catastrophic crop failure. Large areas of single crops are more susceptible to pest and disease attack and also to climatic variations. The impact of drought, for example, is likely to be more serious for a farmer with one type of imported crop than for someone growing a range of different crops, some of which are likely to be drought resistant.

Agricultural changes have caused environmental problems. We now know that agrochemicals often have hidden costs in terms of health and environmental impacts, and that the latter may have direct effects on the medium-term sustainability of farming systems.

Pesticides have direct effects on human health through acute or subacute poisoning. Many pesticides also have known or suspected chronic effects on health, including promotion of cancer and birth defects, neurological effects, and damage to vital organs such as the liver and kidneys (e.g. Watterson 1988).

These effects are far more acute in the South. Many users are unable to afford safety equipment, or even to read instructions (Bull 1982). Advertising is far more aggressive, and often without the safety information that is mandatory in most European countries. Indeed, advertisments sometimes show people using pesticides in hazardous ways, such as spreading granules with their bare hands (Myers 1990). In addition, dangerous pesticides frequently remain in use in Third World countries long after they have been banned in the North, and sometimes Western firms keep making specifically for export an agrochemical that has been banned at home (Weir and Schapiro 1981).

Perhaps more important, in the long term, is the impact that pesticides have on the environment, and on the stability of natural and agricultural systems. Most pesticides also kill natural predators of pests; pests usually recover their numbers faster than their predators so that, when pesticides have been applied once, a farmer is often forced to repeat applications continuously because natural checks have been removed, a situation known as the 'pesticide treadmill' (van den Bosch 1976).

Pesticides also pollute surface and groundwater supplies, damaging fisheries and contaminating drinking water (Bull 1982). The problem of pesticide resistance is becoming increasingly acute, and has already destroyed the effectiveness of many control programmes, including the use of pesticides against malarial mosquitoes in many areas.

Fertilizers are also increasingly seen as a problem for health and the environment. Use of soluble nitrate fertilizers leads to increased nitrate levels in drinking surface and groundwaters, and can also increase nitrate levels in fresh vegetables (Dudley 1989).

Nitrate in water can have implications for both health and environment. High nitrate levels in drinking water can lead to the development of methaemoglobinaemia, or blue baby syndrome, a potentially fatal condition in infants less than three months old. Methaemoglobinaemia is especially likely to occur, and especially hazardous, when water is also contaminated with high levels of bacteria, and where children are already missing vital nutrients such as vitamin C (ascorbic acid), both situations being especially common in the South (Dudley 1989).

In addition, there have been persistent fears that nitrate, acting through the nitrosamines that are one of its breakdown products capable of being formed in the body, could be a causal factor in the development of stomach cancer. Despite much laboratory evidence which appears to support this, epidemiological studies have failed to prove any link (Dudley 1989; Conway and Pretty 1991).

Environmental impacts are mainly connected with the role of fertilizer nitrate in the development of algal blooms in freshwater, which in turn leads to eutrophication, i.e. the depletion of oxygen in water when the algae die and are broken down by bacteria. Here nitrate usually acts in concert with phosphate, another soluble fertilizer. Severe eutrophication can be catastrophic to fish populations.

13

Soil erosion. Intensive farming practices have also increased the chronic rate of soil erosion in many Third World countries (Blaikie 1984). Mechanized farming, which has been the basis of increases in productivity in many temperate regions, is unsuitable for large areas of the world where the soil is poor and conditions arid. Tractors compact soil and break it up, making it susceptible to wind erosion. Tractors and other machines also allow cultivation of a larger area, so that more marginal lands are pulled into production, further increasing erosion (Grainger 1990).

Although these are the best known environmental effects of intensive agriculture, they are by no means the only ones. Amongst others are:

o spillage of organic livestock wastes, leading to eutrophication;
o spillage of silage waste leading to eutrophication;
o spillage of processing waste from plantation crops such as rubber and oil palm, leading to eutrophication;
o release of ammonia from intensive livestock farms and from paddy fields, which is a serious air pollutant in places and can damage growing crops;
o contamination from metals from livestock wastes;
o release of nitrous oxides from fertilizers, which is a greenhouse gas and is also involved in depletion of the ozone layer;
o release of methane from livestock and paddy fields which play a role in global warming;
o biomass burning causes local air pollution and plays a role in acid rain formation, global warming and ozone-layer depletion.

These problems and examples have not been listed in order to argue for a total rejection of all that is modern or innovative in agriculture. However, they do throw some doubts on the claims that current, Western agricultural practices can provide all the answers to the problems of the South.

The impact of the North
One of the new factors in hunger, something that is usually missed out of the simplistic famine equation, is the impact that we ourselves have and, in particular, the role of Western money in fuelling poverty and malnutrition in the South. As in the case of the other broad headings we have been considering here, the impact of the North can be divided into a number of separate categories.

The role of aid agencies and transnational companies
Despite a great deal of rhetoric, most agricultural aid to Third
World countries is still aimed at intensive agriculture, and
still benefits larger farmers over landless peasants or small
subsistence farmers (Lappe and Collins 1976). The structure
of the aid industry, and its intimate links with Western
business interests, means that aid advice is almost always
pushing farmers along similar intensive farming options to
those already being used in the rich countries.

About half of the aid from countries that are members of
the Development Assistance Committee (DAC), which is
connected to the Organisation of Economic Co-operation and
Development (OECD), is tied, i.e. the aid can only be used
to buy goods from the donor country.

One result of this is that pesticides have been heavily
subsidised in many countries, partly as a result of criticism of
the social effects of the Green Revolution (Repetto 1985). The
pesticides may well not be the best for a particular situation
and, as mentioned earlier, are often among the more
hazardous chemicals on the market. Other possible results
could be, for example, that farm machinery is exported to
areas where it can damage soil structure, or that experts are
detailed to oversee the development of large-scale irrigation
projects in places where this may not be the best overall
solution.

This aid bias runs in parallel with the impact of transnational companies (TNCs) in the agribusiness sector, which
are themselves working with business interests in the developing countries. Agribusiness interests amongst TNCs include a
large emphasis on cash crops for export, as these tend to be
the most profitable (e.g. Dinham and Hines 1983).

Involvement by TNCs formerly included a large amount of
direct ownership of land for farming or ranching. Large
agricultural areas of the South are still owned by companies
in the North, and new attempts to buy up land are still
continuing, for example in recent attempts by Cocoa Cola to
fell a large area of rainforest in Belize in order to establish a
citrus plantation. However, today TNCs tend increasingly to
operate in a slightly more hands-off fashion. This can include
part ownership of farming or plantation operations, and also
special deals to buy produce, sales of agrochemical inputs
and other capital-intensive products for agriculture, and,
especially, pressure on countries to grow cash crops for the

15

export market. In this, TNCs work hand in hand with the aid agencies.

The impact of debt Whatever the financial state of a country, there would always be people eager to grow financially rewarding cash crops for export. However, the drive to produce export crops was given an added boost in the late 1980s by the harsh economic conditions imposed by the debt crisis.

The changes imposed by the economic policies of some of the richer nations, and especially the deflationary strategies practised in the United States and UK, known as Reaganomics and Thatcherism, helped create a massive debt problem in the South (George 1988). Third World governments were persuaded to borrow huge amounts during the boom years at the beginning of the 1980s, but were then caught out when interest rates rocketed and global recession reduced their own earning power. Simply servicing the debt proved almost impossible for some countries, and there were fears of a global crash if the largest four debtor nations, Brazil, Argentina, Chile and Mexico, could not or would not continue to pay up (Lever and Hulme 1984).

This had two effects on agriculture. First, it increased massively the pressure on debtor nations to export as much as possible in order to generate dollars or other high-value currencies. Even if governments wished to turn over more land to food for their own people this became increasingly difficult to carry out in practice.

Second, when countries fell into serious trouble they attracted the attention of, and enforced management by, the International Monetary Fund (IMF) which operates on strict and short-term capitalist principles. These have, until recently, favoured intensive, chemically based agriculture rather than traditional or low-input alternatives.

Impact of GATT, CAP and other trading agreements At the present time, there are a number of other factors pertaining to impact from the North. Perhaps the most important is the group of trading agreements that will set patterns of trading over the next decade or more. These include the EC's Common Agricultural Policy; trading links such as those between the African, Caribbean and Pacific (ACP) countries and the EC; the various continental trading blocks operating in the Third World and, perhaps most important of all, the General Agreement on Tariffs and Trade (GATT), currently under renegotiation.

These agreements are dominated by, and provide the greatest benefit to, the rich countries of the North with, at best, similar advantages to a few rich élites in the South. The EC's agricultural support system has disadvantaged exporters from the South for the past two decades.

Current plans to replace this with a far more wide-ranging free trade agreement through GATT are being welcomed by many Third World negotiators, but in practice governments in the South, along with the emerging democracies in Eastern Europe, are likely to lose more from the greater penetration of their economies by Western TNCs than they gain from access to markets in the North (Hines 1991). The environmental impacts of this have already been identified (Arden Clark 1991).

The role of organic agriculture in the South

The main problems with intensive, chemical agriculture in the Third World have been identified as:

o the environmental damage that it causes;
o its attendant unsustainability; and
o the financial constraints that put it out of the reach of most of the poorer farmers and peasants.

The ideal alternative to intensive agriculture would therefore be a farming system that is both cheap and sustainable, and can be operated without causing health or environmental damage. Over the last few years, increasing efforts have been put into developing what have become known as 'sustainable farming policies' for the developing countries.

What is 'sustainable' agriculture?

Unfortunately, one result of this interest is that the term 'sustainable agriculture' has become a much overused slogan and catch-phrase, as different agribusiness interest groups struggle to prove that their own particular products are environmentally benign. They have been able to get away with this because the term has never been clearly defined, and is open to very broad interpretation. (This problem has also beset the definition of 'sustainable development' for the South in general.)

17

Nonetheless, a general picture of the concept of sustainability in farming has emerged. It has come to mean agricultural practices that can be continued indefinitely because they: do not rely on expensive inputs or limited supplies of non-renewable resources; do not cause such gross environmental damage as to threaten the whole future of agriculture in the area; and, in some but by no means all interpretations of sustainability, sustainable agriculture is also defined as farming methods that are suitable for small, mixed farms and for poor farmers as well as for large landowners. In theory, this fits the criteria that were identified earlier. In practice, however, there is still more lip service paid to sustainable agriculture than to substantial changes on the farm.

One result of this is that in farming systems in both South and North, most experiments attempt simply to add 'sustainable' practices onto the existing agricultural system, rather than going back to basics and looking at the agricultural system as a whole. For example, it will be suggested that farmers abandon autumn fertilizer use to reduce leaching caused by heavy winter rains; or that they abandon certain pesticides that have been proven toxic; add a few soil conservation techniques; or increase the area of conservation area on a farm.

The problem is that, introduced piecemeal in this fashion, most of these policies result in reduced production levels without offering anything very substantial in return. Farmers are presented with a confused mixture of different policies and expected to fit them together into a coherent plan for producing food; the new ideas are in consequence often seen as a nuisance and frequently ignored altogether.

There are a number of reasons for this partial approach. Most professional agricultural researchers have been trained in intensive methods and are learning about the new concepts of sustainability as they go along. New techniques tend to get thrown in with existing practices. Many farmers and agricultural policymakers are, not surprisingly, reluctant to throw away all the experiences of the last fifty years (which have resulted in higher yields per unit area in many cases, whatever the other problems) in favour of entirely new approaches. Agribusiness is more than reluctant to see major losses in sales, and is trying to keep farmers using chemicals until new marketable products can be developed, most probably through the use of genetically manipulated organisms (GMOs).

However, more radical approaches do exist to the problem of sustainability in agriculture. Over the last few years, increased attention has been focused on various forms of organic or biological agriculture, which take the idea of changing existing farming practice considerably further, and propose a whole new agricultural system. Far from being a nebulous concept, organic farming is usually practised to comprehensive guidelines, or standards, which define precisely what a farmer or grower can or cannot do within the system.

The definition of organic farming

It has proved difficult to frame a simple definition of organic farming practice, in part because the term embraces a number of slightly different systems operating in various parts of the world. The definition proposed by the United States Department of Agriculture is, whilst not comprehensive, probably a good starting point:

> Organic farming is a production system which avoids or largely excludes the use of synthetically compounded fertilizers, pesticides, growth regulators and livestock feed additives. To the maximum extent feasible, organic farming systems rely on crop rotations, crop residues, animal manures, legumes, green manures, off-farm organic wastes, and aspects of biological pest control to maintain soil productivity and tilth, to supply plant nutrients and to control insects, weeds and other pests.
>
> The concept of the soil as a living system . . . that develops . . . the activities of beneficial organisms . . . is central to this definition. (Quoted in Lampkin 1991)

Proponents of the organic method emphasize the systems approach to farming, i.e. that conversion to organic farming involves getting the whole system in balance, and is fundamentally different from the continuation of chemical agriculture with the addition of a few new techniques to reduce environmental pollution. It concentrates mainly on adjustments within the farm and farming system, and external inputs are minimized, being used as supplements to good management of the internal system (Lampkin 1991).

Organic agriculture has been controversial since the early days of its development during the 1930s and 1940s. Today, it still attracts criticism for being too ideological and for promoting a system that cannot produce sufficient

19

quantities of food. These and other debates will be examined below.

Some misconceptions about organic agriculture

Organic farming has suffered more than its fair share of misconceptions and misunderstandings. Some of these have been created, or at least fostered by those sections of agri-business that do not want to see the success of a system requiring far less of their own services. Nicolas Lampkin identifies four common misconceptions in his major study, *Organic Farming* published in 1991:

'Organic farming avoids all chemicals'

Given that everything is made up of chemical compounds, this is an obvious falsehood. What most organic farming systems do attempt is to avoid the direct and/or routine use of readily soluble, artificially synthesized chemicals for fertilizers, and the use of all pesticides, whether naturally occurring, nature identical, or otherwise. Where it is necessary to use some of these products then the least environmentally disruptive are chosen. Soluble fertilizers are avoided because of their polluting effect, their disruption to growing systems and their impact on soil structure. The arguments against pesticides have been given earlier. Organic farming systems usually avoid all herbicides.

'Organic farming merely involves substituting "organic" inputs for "agrochemical" inputs'

A straight substitution of equal quantities of nitrogen, phosphorus and potassium from manure instead of from mineral fertilizers has little if any advantages from the perspective of either leaching of materials into water courses or soil stability. This approach is not permitted under most organic systems, and has become known as 'neoconventional'.

'Organic agriculture is a return to traditional, pre-war farming'

Although organic farmers and growers use some traditional techniques that have been sidelined or abandoned by intensive farming systems, especially the system of rotation, modern organic practice includes many new techniques and develop-ments, based on contemporary understanding of soil science, biology, ecology, and plant and livestock husbandry. It is a rapidly developing system, where changes in practice, and therefore in the productivity of the system, are occurring continuously in response to research effort and practical on-farm experience.

'Organic farming involves a fundamental lifestyle change'
The basis of much Western suspicion of organic farming is that it is rooted in extreme political, philosophical or social movements that are in opposition to the rest of society. Most contemporary organic practitioners are commercial farmers who have converted to an organic system either because they are unhappy about the practices or the sustainability of conventional farming and/or because they see organic food as offering a new market for their produce. In addition, a few biodynamic farmers follow the wider philosophy and teaching of Rudolph Steiner, the German philosopher who invented the system (a specialized form of organic agriculture). There is no coherence to a particular political perspective within the organic movement.

The practice of organic agriculture
Organic agriculture has been gaining support steadily in the North, although it would be disingenuous to claim that opposition had evaporated. A steady increase in the number of organic farmers has been matched by political and academic support, and increasing credibility amongst farmers.

Support from research bodies   An influential report from the US National Academy of Sciences found that pollution from organic farming systems was far lower than in the case of conventional farming and, significantly, that there was relatively little overall decrease in productivity from a switch between the two:

> [reduced use of agrochemical inputs] lowers production costs and lessens agriculture's potential for adverse environmental and health effects without necessarily decreasing, and in some cases increasing, per acre crop yields and the productivity of livestock management systems. (NAS 1989)

Research in the UK suggests that yield reductions under an organic system can vary between 10–30 per cent at the current state of development of organic agriculture (Vine and Bateman 1981). Comparisons at Kassel University in Germany found average reductions of 10–15 per cent (Vogtmann 1990).

21

Direct comparisons of yield are notoriously difficult, and depend at least in part on the degree of intensification being practised nearby. Organic farming will probably never be able to compete grain for grain with the most intensive wheat production, but the latter is having to be cut back in any case because of environmental side-effects. Further complications are created when the quality of food is compared. Crops that have been rapidly boosted with soluble fertilizers can sometimes contain increased levels of water, and decreased concentrations in valuable nutrients such as vitamin C (Lampkin 1990), so that a direct comparison of weight does not give a full picture of nutritional benefits.

In temperate regions, there have been a few preliminary attempts to analyse what a larger-scale shift to organic agriculture would mean for food supply. Lampkin and Midmore (1989) used a computer model to predict the impact of a 20 per cent shift to organic methods in Britain, in response to the Soil Association's *20 per cent Organic by the Year 2000* campaign, and found that there would be virtually no effect on nutritional output, although some types of produce would increase and others decrease slightly. As yet, no similar exercise has been carried out in any developing country.

Political support   Organic farming has also started to receive political support in Europe and North America, with limited government funding in some countries, such as Denmark and the Netherlands, and backing from official and non-governmental organizations that have traditionally been suspicious of organic practice.

In the UK, for example, the government's Nature Conservancy Council (now divided into three regional bodies) stated its support for the Soil Association's campaign to convert 20 per cent of Britain's farmland to organic methods by the end of the century (NCC 1990). Most environmental organizations back organic agriculture. The four main political parties, Conservative, Labour, Liberal Democrats and Greens, have all stated their support for organic agriculture in principle although in the Conservative government's case this is as yet more in words than deeds.

Repeated consumer surveys suggest that most British consumers would like to have greater access to organic food, and support organic farming. A survey organized by the *Daily Telegraph* newspaper suggested that most people would prefer organic farming to options currently set aside being used to reduce overproduction in the EC.

Despite the widespread support for organic agriculture, organic farmers still farm only from 0.5–3 per cent of agricultural land in European countries. At least part of the reason for this is that the structure of European agricultural funding, through the Common Agricultural Policy, acts as a deterrent, as does continued, entrenched opposition from sectors of the farming and chemical industries.

In the South, there is still even greater suspicion, and frequently claims that promotion of organic methods would be counter to the general good, because maximum production of food is required. We consider this to be an oversimplification, and we will now look at some of the advantages presented by the organic system in the Third World.

The advantages of an organic system in the South

The merits of organic agriculture have been assessed from three perspectives: agronomic, environmental and social. There is a great deal of overlap between the three.

Use of rotations: i.e. growing different crops on a piece of land in a planned rotation over several years. Rotations help maintain soil fertility and soil structure without adding high inputs of soluble fertilizers. Most rotations include a leguminous crop, that is a member of the pea or bean family, because these contain symbiotic bacteria in their root nodules which are capable of 'fixing' nitrogen from the air; after harvest the plant remains can be ploughed in and the soil nitrogen levels raised as a result.

Rotations also help break pest cycles. Pest-specific crops can remain in the soil from one year to the next, often over-wintering as larvae or eggs, but the several-year gap between planting a particular crop in rotation usually ensures that the pests have disappeared by the time it is planted in the same plot again (Blake 1989). Thus rotations reduce reliance on bought-in inputs and increase crop diversity, which fits in better with subsistence or semi-subsistence farming. Rotations can be practised on a farm or garden scale.

Intercropping or mixed cropping: i.e. growing two or more crops together simultaneously. This can help make maximum use of available space because the plants often use different nutrients and, in the case of legumes, can actually increase levels of soil nitrogen. Crops can also interact beneficially; for example a fast-growing crop can provide protection for one which germinates slightly later, or a strongly scented plant can help disguise another which is subject to attack by pests that locate it by its smell.

Intercropping can prevent soil erosion, by ensuring a continuous vegetation cover over soil, and has the social advantages of cutting down on weeding because of the density of crops and ensuring that the work of harvesting comes in manageable portions throughout the year rather than all at the same time, as in the case of a monocrop (Jackson 1990).

One particular form of intercropping is agroforestry, that is, the growing of farm crops and trees together. This can be a way of maximizing efficiency of land-use in forestry plantations, so that farmers intercrop trees and crops during the first few years when the trees remain small; an example of this is the Tanguy system practised in parts of northern Thailand (Myers 1984). In other areas, farmers have developed methods of growing crops in between mature trees in forests, thus allowing farming to continue without causing deforestation. Intercropping is a traditional practice in many Third World countries. It is now being replaced by monocrop systems.

Green manuring: the practice of growing a legume, or other nitrogen fixing crop, in order to plough it in and increase soil fertility. Commonly used green manures are *Sesbania* and *Azola*, a fern containing blue green algae that can fix nitrogen (Conway and Barbier 1990). Green manures are useful both as a way of maintaining soil cover, because soil is protected from erosion during periods of the year when crops are not being grown, and is also a way of improving soil fertility without recourse to expensive agrochemicals.

Prioritizing soil conservation: In many Third World countries, in arid areas or where deforestation has increased flooding and climatic instability, this is now probably the single most important farming technique (Eckholm and Brown 1977).

Organic methods have been shown to benefit soil conservation in a number of different ways. Comparison of a long-term organic farm with a neighbouring chemical farm in the United States found a humus layer sixteen times greater in the former (*Living Earth* 1989). Earthworm densities have been shown to increase on organic farms (EFRC forthcoming), and the benefits of earthworms in maintaining soil structure has been clearly demonstrated (e.g. Stewart *et al.* 1988). In addition, the emphasis on small field sizes in an organic system allows other soil conservation methods, such as planting shelterbeds, dykes and use of drainage channels, to be introduced into the system more easily than if farmers are relying on large fields.

Non-chemical pest control: organic methods minimize the use of pesticides and recommend a range of other pest-control techniques. These include: the encouragement of natural predators, by leaving sufficient areas of natural and semi-natural vegetation for them to survive; the occasional and carefully monitored application of biological controls, such as the bacteria *Bacillus thuringiensis*, against some insect larvae; use of traps and barriers; mechanized weeding; and a range of cultural controls, including timing planting to avoid pest cycles, and intercropping (Blake 1988).

These methods have clear advantages from health and environmental perspectives; they also relieve poor farmers and sharecroppers of the financial burden of buying high-cost inputs. If applied well, they also help prevent the regular boom and bust cycle of pest attack caused by regular use of pesticides, and help to alleviate the mounting problems caused by pesticide resistance.

Low input of agrochemicals: including both pesticides, mentioned above, and artificial fertilizers. The latter are replaced by cultural techniques, including rotations, intercropping and use of green manures, and from managed use of composts and animal or human manures where these are available. This has advantages for human health amongst workers and others drinking water from the farming catchment, and helps prevent damage to local fisheries and to wildlife.

Recycling: the emphasis on management within the farming system, rather than importing nutrients and control methods from outside, means that recycling is given a priority. This includes composting plant remains, agricultural wastes and manures. It has a number of advantages over use of soluble fertilizers, in terms of both costs and the quality of soil structure produced. However, this can present some problems in areas of the South where agricultural wastes and manures are used as fuel. Adequate recycling will only be possible if the energy problems of the area are also tackled, for example, through fuelwood plantations, use of more efficient stoves and solar cookers.

Low degree of mechanization: although organic farming can, in principle, be practised on a large and highly mechanized scale, it is far better suited to a low-mechanization approach than is the case with intensive chemical agriculture (McRobie 1990). This reduces the disadvantages currently faced by small farmers and those on marginal land, where mechanization is not possible or affordable. Reducing or eliminating use of heavy tractors also lessens the impact on soil structure and helps to prevent erosion (Grainger 1990). From a social perspective, mechanization is less important to efficient production in countries where there is often an excess labour force and little money to buy expensive machinery.

Mixed farming: the practice of growing a range of crops, and of mixing livestock and arable farming, rather than concentrating on one or a few types of produce. The use of rotations ensures that organic farming remains mixed. Research in Germany has found that the average range of types of produce often triples when a farm converts to organic (Vogtmann 1990).

This helps farmers who are practising subsistence farming and growing crops to sell at the same time, and is also generally more beneficial for soil structure and fertility than the monoculture approach. Mixed farming tends to reduce pest attack and allows flexibility to include local crops, which are often forced out when farming is modernized and restricted to the priorities of Western agriculture.

Suitable for small areas: many of the techniques and advantages listed above mean that organic methods are better suited for small rather than large farms. This is of key political importance, as will be discussed in the next section, and is an important reversal of trends in agriculture that have favoured the larger and more powerful landowner above the peasant and smallholder.

Builds on existing skills: although we have stressed that organic farming is a new technique, it builds on many existing skills likely to be found among peasant farmers everywhere. (The case of nomadic herders, who are now increasingly being forced to become settled farmers, is different and requires special treatment, whatever farming system is being practised.) Organic agriculture is also as accessible to women and children, who perform most of the traditional agricultural practices in many areas, as it is to men. Indeed, some of the techniques used in traditional agriculture have proved a useful starting point for the development of contemporary organic techniques.

Indeed, the erosion of peasant knowledge is now regarded as a serious problem. Efforts are being made to collect and use traditional techniques, such as that organized by the Agroecological Project of Cochabamba in Bolivia (Rist 1991).

Low energy use: the energy crisis facing many Third World countries is the most critical of all energy issues today. Over-reliance on machinery and on inputs requiring high energy use in their production (such as artificial fertilizers) puts farmers at risk from price fluctuations or real shortages. Research in Germany has shown that average energy use on an organic farm is less than on a conventional farm (Vogtmann 1990); this is likely to be even more so in the South where mechanization is limited and farm size is small. For many farmers, where labour is plentiful and capital in short supply, simple tools are far more cost-effective than expensive machinery (McRobie 1990).

Reduced environmental impact: it is becoming increasingly clear that concern about environmental protection is not something that is confined to the middle classes in the North. Issues such as groundwater and freshwater contamination, deforestation and soil erosion have become of critical importance to millions of people living in the South. A farming system that prioritizes ways of reducing these impacts has obvious advantages.

Reduced health impacts: this is also true for human health. The problems of pesticide abuse are now widely recognized by groups and individuals in the South. Farmers believe that they are hooked onto the use of agrochemicals, whilst often

27

being fully aware of their problems, as the following quotation from a series of case studies collected by the Farmers' Assistance Board of the Philippines illustrates: 'I have to continue using pesticides or else farm yield would be insufficient for my family's subsistence. But every time I spray pesticides, I experience nausea, headache, backpain, diarrhoea' (FAB 1982).

Farming methods that offer a way out of this impasse will receive considerable support from the farmers and farm workers who are currently suffering the ill-effects of agrochemicals.

Organic agriculture thus offers a number of substantial advantages over a conventional system, in terms of environmental stability, efficiency of resource use, producing food for human needs and matching existing skills to improved production techniques.

Putting organic farming into perspective in the South

There are, of course, also some limitations. Organic agriculture is not a system that can simply be duplicated, unaltered, anywhere in the world. Much research and development is needed to develop organic systems suitable for local conditions and, at present, such specific and directed research does not fit comfortably into the structure of aid agencies, or indeed the agricultural research programmes of many countries in the South.

In addition, any low-input techniques will continue to meet entrenched opposition from agrochemical and agribusiness interests. Although major financial institutions, such as the World Bank, are now paying increasing lip service to low-input agriculture, this has yet to be translated into any very specific change in direction of policy.

One of the problems is that we still know little about comparative productivity of different systems in the South and, given the totally different mixture of crops that will be grown, a direct comparison will probably remain difficult. We do not know, as yet, whether it will be possible to introduce a completely organic system in some of the most badly damaged environments, or whether it will be necessary to, for example, use soluble nitrate fertilizers to restore fertility fast enough to feed people currently reliant on nothing but degraded land or semi-desert. There is an urgent need for further research into organic agriculture as it will be applied to areas of the tropics. This is now starting, albeit slowly and often underfunded.

Another major problem is the difficulty of getting information across to farmers in the field. Intensive agriculture is promoted by the efforts of sales representatives, high-profile advertising campaigns and, to a very considerable extent in the past, by at least some aid and development agencies. Organic farming, on the other hand, relies mainly on small projects, volunteers and individual enthusiasts. Poor communications and transport make spreading information very difficult, and the high-tech solution often appears more plausible.

In the mid-1980s, farmers and teachers in Tanzania complained to me that there was no information available on composting, yet the US-based Rodale Institute had spent considerable funds in producing basic literature in Swahili and distributing this in East Africa. Without the kind of professional infrastructure available to TNCs or the largest aid agencies, spreading information about organic farming or any other appropriate technology development is a difficult and time-consuming process.

Case studies of organic agriculture in the South

Notwithstanding the limitations outlined above, interest in organic agriculture is far from being confined to a middle-class élite. Current membership of IFOAM, the International Federation of Organic Agriculture Movements, includes more groups in the South than in Europe and North America; it also has a rapidly expanding east European membership. In this section, we give some examples of organic farming systems already being practised successfully in the Third World.

Case Study 1: Organic rice production in the Philippines

Mamerto Fantilanan, a former postmaster in his sixties, has twice been named Philippine Farmer of the Year for his work on a half-hectare fully organic rice farm near Cuartero in Capiz. Nutrients are supplied by growing the fern *Azolla*, which contains a nitrogen-fixing blue green algae, *Anabaena azollae*, amongst the rice plants. When it dies and is incorporated into the soil, decomposition frees the nitrogen. A rice crop fertilized with *Azolla* can yield about 1.5 tons a hectare more than an unfertilized crop.

In addition, *Azolla* is used along with manure to power a small methane digester, and the sludge is in turn used to fertilize rice, vegetables, root crops and fruit-tree seedlings. Pests are controlled by a range of integrated pest management techniques. In recognition of the system's success, the FAO and the Capiz provincial government built a training centre at the farm in 1987, which trained over 300 farmers in the first two years of its existence. Experiments with *Azolla* were first carried out by the Manila office of the International Rice Research Institute (IRRI 1989).

Case Study 2: Institute for Sustainable Agriculture, Nepal (INSAN) INSAN is a Kathmandu-based institute based on the principles of permaculture, a system that goes considerably beyond the principles of organic farming in developing detailed site designs which take into account all aspects of food production and energy needs. INSAN is establishing a network of model farms throughout Nepal to develop the principles within the country. To date the institute has two of its own farms and six associate farms run by INSAN members.

All the farms operate as model demonstration sites, conduct extension outreach programmes, offer consultancy services and provide seeds, saplings and other material to interested farmers. On-farm research is also carried out. The INSAN Sunsari farm project is investigating the potential of establishing Neem tree plantations to use the oil as a natural pesticide. Research is being funded by the World Bank to investigate use of Vetiver grass for erosion control. Experiments on no-tillage farming are taking place in Western Nepal. INSAN also hosted the Fourth International Permaculture Conference in February 1991 (INSAN 1991).

Case Study 3: Manor House Agricultural Centre, Kenya Manor House Agricultural Centre was established in 1984 to help small-scale farmers in Kenya by spreading information about biological, or biointensive, agriculture 'to communities that are unable to afford high-cost external inputs as inorganic fertilizers and pesticides' (Manor House 1990).

It operates mainly by teaching community-sponsored education trainers the methods of biological agriculture, through detailed two-year courses, the first of which began in 1986. Teaching includes composting methods suitable for local conditions, rotations, companion planting and use of a diverse mix of crops. Manor House also teaches small-scale livestock raising and use of animals for traction.

Many of the students are sponsored by village communities, which are then encouraged to practise the techniques in their own farms and plots. The centre also has extensive gardens, where organic agriculture is carried out, and the results monitored (Manor House 1988).

Case Study 4: Ambangulul Tea Estate, Tanzania One of two major tea estates that successfully applied for the Soil Association Symbol to export organic tea to the UK in 1990, Ambangulal lies at 3500 feet in a cool environment, on steep slopes. Tea plantations are interspersed with remnant forest and eucalyptus plantations to provide timber for tea drying. Plots of maize, vegetables and bananas are also grown for food by the workers.

Initially only part of the estate is being converted to an organic system. Pests and diseases are not usually a problem due to the high altitude; however some use of nitrogen fertilizer will have to be replaced with manures or other acceptable organic alternatives (Dalby 1990).

Case Study 5: Development of the Zai technique in Burkina Faso Zai is a traditional agricultural method, traditionally used in Burkina Faso, to cultivate and redevelop the arid and crusted part of fields that have suffered from severe erosion. Following widespread desertification and erosion, the Zai method is being reintroduced by the Yatenga people, with the help of the Ministry of Peasant Co-operative Action.

The method consists of three main stages: making small holes in the arid earth, filling them with compost and sowing into the composted earth. Seed holes are 20–30 cm wide and 15–20 cm deep, and are made at a density of 20,000–25,000/ha for sorghum and millet. About 5–7 tonnes/ha of compost are required. The Zai method is effective for these conditions because it helps concentrate water in specific areas where it does the most good, keeping compost moist and helping the crops grow. Results are about 900–1200kg/ha, a yield increase of 30–35 per cent. The success of the Zai technique has encouraged the ministry to set up experiments and information exchange between peasant groups (Ouedraogo 1990).

Case Study 6: Organic farming techniques for sugarcane plants in the Philippines The Institute for Agricultural Systems and Solar Resources at the University at Los Banos has been researching ways of reducing fertilizer applications on sugarcane plantations. Although a fully organic system has not yet been achieved, initial research suggests that fertilizer use could be halved without yield losses, even on the large and highly mechanized estates.

Two methods were investigated:

○ *Trash farming*   The conventional equidistant row spacing is replaced by double rows, where two rows of canes are planted nearer together than in a conventional system but a larger gap is left before the next two rows. In this gap, trash and tops from the sugarcane can be left to decompose, rather than being burnt as in a conventional system. This replaces much of the nutrient usually supplied by inorganic fertilizers.

○ *Intercropping with a rhizobium-inoculated legume*   Use of an intercropped legume, capable of fixing atmospheric nitrogen was also found to increase yield in the second year (ratoon) crop.

The researchers noted that although smaller holdings were more suited to an organic method than the largest plantations, the uncertainty of land tenure was such that those farmers 'do not find the practice of building/restoring soil fertility attractive' (Mendosa 1990).

Principles and practices of organic agriculture

The International Federation of Organic Agriculture Movements (IFOAM) acts as an umbrella group for many national organic organizations. It lays down basic standards that all member bodies must meet. These include a concise expression of the principles and practices of organic farming, and are reproduced below:

○ to produce food of high nutritional quality in sufficient quantity;
○ to work with natural systems rather than seeking to dominate them;
○ to encourage and enhance biological cycles within the farming system, involving micro-organisms, soil flora and fauna, plants and animals;

32

o to maintain and increase the long-term fertility of soils;
o to use as far as possible renewable resources in locally organized agricultural systems;
o to work as much as possible within a closed system with regard to organic matter and nutrient elements;
o to give all livestock conditions of life that allow them to behave naturally;
o to avoid all forms of pollution that may result from agricultural techniques;
o to maintain the genetic diversity of the agricultural system and its surroundings, including the protection of plant and wildlife habitats;
o to allow agricultural producers an adequate return and satisfaction from their work including a safe working environment;
o to consider the wider social and ecological impact of the farming system.

The role of land reform in the South

At the present time, the bulk of good agricultural land is owned and/or controlled by a small minority of rich landowners. This general pattern is true for the whole of the non-socialist world, but is especially pronounced in the South. Some statistics illustrate the size of the disparity:

o in Peru, 93 per cent of agricultural land is owned by the top 10 per cent of landowners (Seagar 1990);
o in El Salvador, 78 per cent of the land is owned by the top 10 per cent of landowners (Seagar 1990);
o in Brazil, 2 per cent of the landowners own 60 per cent of the arable land and 70 per cent of rural households own no land at all (Holmberg *et al.* 1991);
o in the Philippines, 4.2 per cent of the population owned all the land before the overthrow of President Marcos (Farmers' World Network 1991);
o even in countries that still claim to be socialist, such as China, there has been a gradual reversal back towards private ownership (Roeleveld, 1987).

The United Nations FAO estimates that 30 million rural households worldwide own no land at all, and some 138 million households are almost landless, two-thirds of them in Asia (Jodha 1990). Many of these people are tenants, or illegal

33

squatters on marginal land, in once-forested areas or in deserts and semi-deserts.

The causes of landlessness
The extreme nature of landlessness in the South has been caused by a wide range of problems, perhaps especially common in developing countries.

Political unrest Many Third World countries gained their independence from colonial powers only after prolonged political unrest and, sometimes, a protracted civil war. Others have had external or internal wars through uprisings by the right and left, tribal disagreements and nationalist wars. All of these periods of unrest, religious conflict and political upheaval provide opportunities for the rich or ruthless to increase their control over land areas.

The Green Revolution As already discussed, the increased economic disparities created by the Green Revolution allowed larger landowners to gain control over increased areas of arable land in a number of countries.

Population increase The rapid increase in population over the past few decades has thrown strains on land space in a number of countries. In Thailand, for example, the rural population has more than doubled since 1940, despite large-scale migration to urban areas. Kenya's population, which is doubling every twenty-five years, is far outpacing the creation of new farmland (Eckholm 1979).

Underpricing of land One reason why additional land is brought into production, often very inefficiently, is because non-agricultural areas are not costed in financial terms. This means, for example, that it is worth burning or otherwise clearing forest areas for farming, even if the productivity is limited and the land only remains worth farming for a few years.

The result is a steadily increasing imbalance; at the very time when population growth means that a more equitable distribution of land is needed, the ownership of good farmland has been passing into the hands of fewer and fewer people. Erik Eckholm, of the Worldwatch Institute, summarized the problem in a paper published in 1979:

Whilst rural land tenure and social patterns vary greatly from place to place, it is generally true that where a few individuals own a large share of the land, these same individuals dominate local politics and, through their roles as lenders, landlords and employers, the economic lives of their neighbours. In other regions, a larger number of farmers owning small or medium-sized plots may predominate. Under such conditions these landowners, too, may be the controllers of wealth and power; at the least, they usually enjoy a certain economic security and the possibility of personal economic progress.

Whatever land tenure pattern prevails in a given area, it is the landless and near landless who are at the bottom. Hundreds of millions of families are struggling to improve their lives through agriculture without secure access to the basis of agricultural life, farmland.

(Eckholm 1979)

The impact of landlessness

The landlessness problem is not just something that affects people's wealth or security; it also has direct effects on whether they live or die in times of food shortage, and on the environmental stability of the area. A number of critical land-tenure issues are examined below.

Landless people have no land to produce food   This apparent truism is worth stating as the first 'impact' of landlessness. Although many landless people are tenants or work on the land, others have virtually no access to land, and all are most severely at risk in times of drought, flood, war or other causes of food shortages. In Bangladesh during the food shortages of 1975, the death rate amongst landless people was triple that of people owning three or more acres of land (Brown 1976).

Land that is available is overused   As families expand in size, and available land contracts, farmland is overexploited causing loss of fertility, soil erosion and, if the family can afford them, over-use of agrochemicals. For example, the migration of rice farmers into the Northern uplands of Thailand has resulted in a shortening of the previously stable slash-and-burn system from a cycle of 8–10 years down to between 2–4 years in many places. This increases erosion, leading to increased flooding and siltation of dams (Brown and Wolf 1984).

35

Forests, and other 'wild' areas, are cleared for farmland   The relatively weak controls over, or even active encouragement of, settlement of forests and other wild areas results in landless peasants moving into forests on a large scale throughout Asia, Latin America and parts of Africa. Apart from the global impacts on biodiversity and climatic stability, clearance of forests has a number of direct effects on other farmers in the area, by: increasing soil erosion and flooding further downstream; changing local weather patterns, including the possibility of increased drought; and destroying a common resource of wood fuel, nuts, berries, wild food plants and game, which can provide up to 50 per cent of families' resources in some areas.

In addition, the soil under many tropical forests is unsuitable for agriculture, resulting in land being farmed for a few years and then abandoned to secondary scrub or even desert (Myers 1979).

Farmers are forced onto more marginal land   It is not only in the once-forested areas that landless farmers are forced to try and eke out a living. In arid regions of, especially, Africa and Asia, marginal and near landless farmers increase environmental stress on lands that need to be used with extreme care, if at all. The growth of a market economy has, for example, resulted in previously nomadic herders being forced into settled farming in unsuitable areas, and without them having the necessary skills or history to enable them to use the land efficiently (Grainger 1990). In many areas, large landowners have expropriated or otherwise obtained title to previously common land (Holmberg *et al.* 1991).

Landless tenants have no reason to use the land sustainably   A major problem with any attempts to decrease environmental damage is that people without any security of land tenure have little incentive to look after the long-term ecological interests of land that they are renting. It also means that the owners of the land often live far away, and have little idea about conditions there, or any innate feel for farming the land with future needs in mind. Instead, they want quick returns on their investment, which often means flogging the land as hard as possible. This problem of land tenure was identified in the United States during the 1950s:

Many farms in the Corn Belt are owned by absentee landlords who have little personal contact with their tenants. These owners do not realize that conservation adjustments will improve farm income over a number of years. Instead, they want a high return on their investment now. On many farms the tenant is also interested in short-run profits. He may have only a one-year lease with no assurance of renewal, or the leasing agreement may require him to shoulder a larger share of the conservation costs than he receives in benefits. (Quoted in Eckholm 1979)

Indeed, in some areas of India, landowners do not allow tenants to farm the same land from year to year for fear that they may lay claim over the land (Walinsky 1977).

Lack of tenure keeps land under cultivation   One specific reason for over-use of land by tenant farmers is that they fear that leaving land fallow will result in them being dispossessed in favour of others who will use it more intensively. Sustainable practice, on the other hand, would favour occasionally leaving the land fallow to recover soil fertility and reduce the rate of erosion (Holmberg *et al.* 1991).

Land use is more inefficient   We have already touched on some of the causes of inefficiency above: small farmers being forced onto marginal land and into unsustainable farming. There is also evidence that, especially in the South, small farms are more efficient than larger holdings in terms of output.

A joint World Bank and International Labour Organisation study found that yields per hectare were broadly comparable in similar areas of large and small farms, but that small farmers tend to use their land more intensively, planting a greater share of it and double-cropping wherever possible. A detailed study of Brazil, Colombia, India, Malaysia, Pakistan and the Philippines concluded that a transition in those countries to uniformly small family farms could increase national agricultural output from amounts ranging from 19 per cent in India to 49 per cent in Pakistan (Berry and Cline 1976).

Conclusions   Serious inequalities in land distribution are bad for most of the people and for the environmental stability of an area, especially in countries where a large proportion of the population is involved in agriculture. They encourage damaging land-use patterns, and the clearance and settlement of unsuitable land. They also reduce the incentive for people to look after land efficiently and carefully. Far from being simply a major social issue, inequality is also a fundamental environmental one, one that has a direct relationship with the ability to feed a country's population. Land reform, the legally implemented redistribution of land to a greater number of people, is thus one of the key issues that needs to be addressed by food policymakers and activists.

Giving people back their land
This is easy to say, but far more difficult to achieve in practice. Countless governments have been elected in both Asia and Latin America on the ticket of sweeping land-reform policies, only to fail in their stated intentions under an onslaught of vested interests, corruption, political inefficiency and, more than once, military takeover.

Land reform is both politically and practically difficult, and has seldom been achieved in any radical manner without violent revolution, as in the examples of China, Cuba and the eastern European states. In many cases, other inefficiencies in these political systems meant that the land reform itself was not sufficient to increase net agricultural productivity. Other land reform legislation has been rushed through by conservative governments in the aftermath of war and when in fear of further unrest; examples of these pressurized reforms include Japan, South Korea and Taiwan, which all instituted land reform to hold off the possibility of communist uprising (Eckholm 1979).

In many other cases, there have been unsuccessful attempts to promote land reform by violent means. In 1932, almost 20,000 El Salvadorean peasants lost their lives in fighting for land. Similar battles have been fought in Guatemala and indeed most other Latin American states (Eckholm 1979).

Right up until the present time, most aid agencies and donor governments have dodged the fundamental issue of land reform. This is partly because the rich élites of North and South collude with each other to keep landlessness off the political agenda, and partly because fieldworkers cannot be too outspoken about land rights issues if they wish to remain in the country at all.

Instead, a number of softer reformist initiatives have been launched, including the community development movement

UNIVERSITY OF WINNIPEG. 515 Portage Ave., Winnipeg, MB. R3B 2E9 Canada

of the 1960s, which was aimed to bring rural people together to work for shared goals. This faltered, and eventually collapsed, against ingrained inequalities, including those of landownership, exploitation by élites and urban domination which 'could neither be ignored nor bypassed' (Holdcraft 1978).

Some of these made things worse rather than better. For example, one Swedish-funded programme to boost crop production amongst poorer tenant farmers was so successful in agronomic terms that the landowners evicted thousands of tenants to begin farming themselves, and raised rents for the rest (Cohen 1978). Nonetheless, the critical importance of land reform means that efforts to achieve it are still continuing.

Case studies of land reform

Despite intense political pressure for over 30 years, the majority of land remains in the hands of a tiny élite. On 13 March 1964, newly elected President Goulart announced that he would begin an agrarian reform programme, but was overthrown by the military within a fortnight, leading to 21 years of authoritarian rule.

Nonetheless, the popularity of the initiative had been so immense that some efforts were made to pay lip-service to the principles. Laws were introduced to distribute the underused *latifundios* and redistribute them amongst squatters, share-croppers and landless labourers. A Territorial Land Tax was also introduced to tax unproductive land and use the money to help finance redistribution.

However, large landowners lobbied hard to water down the proposals which, in any case, remained largely unimplemented. President Medici (1969–74) emphasized colonization rather than reform, and started the great (and largely unsuccessful) colonization of the interior Amazon basin. President Giesel (1975–79) emphasized improvement in agricultural productivity rather than land reform.

By August 1985, more than US$2b of back tax was owing from the Territorial Land Tax, and the whole process had lost credibility. In the previous five years, at least 787 people had been killed in conflicts over land; many more deaths were doubtless unrecorded.

President Sarney, who was in power until 1990, attempted to revive the original proposals, but again nothing happened and this was a major factor in his loss of office. The new presidency has begun with still-unresolved conflicts, and landownership remains polarized between the rich élite and the vast landless majority (*Food Matters* 1991).

The Philippines From the late eighteenth century, many monastic orders and private landowners started to rent out areas of land to peasants. Following US occupation, a newly rich class of landowners appeared, obligations to tenants were said to be less carefully addressed than before, and political opposition to landlessness increased. When Ferdinand Marcos became president he promised to make 350,750 share tenants into leasees, but by 1970 only 8 per cent of this small target had been reached.

In 1971, faced with growing calls for agrarian reform, President Marcos imposed martial law. The Department of Agrarian Reform was created and Marcos declared the whole country to be an agrarian reform zone, with tenancy to be illegal. Despite much talk, nothing material occurred and this was a major factor in the 'snap revolution' which brought Corazen Aquino to power in 1986.

Aquino, once again, swore that agrarian reform would be a cornerstone of the new society. However, with 75 per cent of the parliament being major landowners, implementing these objectives has proved an elusive goal (*Food Matters* 1991). About 50 per cent of the population remain tenants, much land is now in the hands of TNCs, leaving 85 per cent of the farmers growing on marginal hill slopes (Farmers World Network 1991 and CIIR 1987). An FAO report concluded: 'The reach of the agrarian reform has barely touched the Filipino peasant'.

Zimbabwe Following independence in 1979 there have been considerable efforts at land reform by the government. These have had some success, but have also thrown up a number of problems.

Land reform has followed four basic models:

o Farms bought from European settlements, which were divided into zones with some fifty families, each having 5 hectares for arable production and grazing rights of a size which is dependent on the productivity of the land. There has also been development of clinics, schools etc, and this initiative accounted for 85 per cent of the settlement.

40

- Attempts have been made to develop co-operative tea and coffee estates, but these have not been particularly successful.
- Some estates are run by out-growers, who then hire facilities for processing off the estate.
- Attempts were made to settle people on purchased range-land. However, land is centrally managed for a grazing system and crop production is not allowed; again this system has not proved very successful.

Some 48,000 families have been settled on 3m ha of land, but many settlers were too poor to use the land efficiently, while others have been overstocking their areas. There are still questions about whether land is passed on to children and, indeed, about how secure present tenants are. The environmental impact of the scheme has yet to be thoroughly assessed (Farmers' World Network 1991). Nonetheless, this land distribution has gone far further than the two previous examples, or than most peaceful attempts at land reform.

Despite years of efforts at achieving land reform, this has not occurred in significant ways in most Third World countries. Yet the chances of implementing policies capable of feeding the growing populations of these countries, and maintaining a sustainable agricultural system and stable environment, seem increasingly to rely on some form of redistribution of land. Support for those groups working towards this aim and, perhaps, pressure on rich élites within countries not to stand in the way of change is seen increasingly as an important part of the agenda of many development groups.

Sustainable farming and land reform

The conference which took place in Berlin in November 1991 was a first conscious attempt to increase liaison between the land reform and organic movements on an international scale. Such an initiative is particularly important at the present time. The problems facing the poorest people are likely to increase over the next few years. Changes in international trade laws, and the revision of the General Agreement on Tariffs and Trade (GATT) will give an additional trading advantage to the large companies of the North and the rich élites of the South.

Other changes threaten the poorest farmers. Developments in genetic engineering look set to further reduce the value of

41

at least some of the agricultural cash crops of the South. For example, it will soon be possible to synthesize the insecticide pyrethrum in the laboratory, instead of importing it from small growers in Kenya. The increasing shift towards patenting life forms and intellectual property will mean that any fresh developments in crop breeding will only be accessible to the richer farmers.

It is important to stress once again that solutions must be pragmatic rather than ideological. We don't know if a fully organic system is necessarily the best option for degraded land in drought-stricken areas, for example. There may well be uses for soluble fertilizers in some cases, perhaps to help restore fertility to impoverished soils. The important principle is that any agricultural approach should be a systems approach and should minimize inputs that are expensive and/or hazardous to health, or environmentally disruptive.

Of course, other priorities are important for developing sustainable food policies, including education, a changing role for women, primary health care, social security and, above all, a change in the economic relationships between the South and North. None of these changes will be possible without a balanced agriculture and an adequate food supply. Achieving such a goal will involve active co-operation between the many groups and individuals all working towards the same basic aims, from different starting points.

Concept, Requirements and Measures to Ensure Sustainable Agricultural Development

UWE OTZEN

Scale and causes of the destruction of agricultural resources

The growing destabilization and destruction of agricultural resources is a part of the worldwide ecological crisis. It is characterized by a wide range of environmentally damaging factors, which have many causes. Directly or indirectly, various of these factors are in turn linked to create further problems, which adversely affect the assimilative and regenerative forces of natural resources. The resulting interdependence stems from a mixture of economic structures, technology, demographic trends, climatic factors, conflicts and disasters, social structures and also simply from public attitudes towards the environment. In the agricultural sector we can distinguish the following key areas, which are all, ominously, linked:

o the exploitation of timber resources for exports of raw materials and the environmentally incompatible clearance of tropical rainforest for use as pastureland, plantations or arable farming. This initiates the normally irreversible destruction of rainforest ecosystems, which are of value to the whole of mankind and contain an enormous wealth of species;
o increasing levels of unchecked and now dysfunctional shifting cultivation in humid and dry savanna areas, causing the rapid destabilization of sensitive savanna ecosystems;
o large-scale monocrop agriculture in dry and rain-fed areas that once featured mixed peasant farming, eventually leading to soil degradation in fragile savanna ecosystems;
o the spread of arable farming into increasingly marginal savanna areas and mountainous regions, thus initiating soil erosion and overuse, and potentially leading to the complete destruction of the ecological balance in these zones;
o the spread of intensive grazing, encouraged by irrigation programmes in zones which have hitherto been unused or only used under a very extensive system. This results in

43

the overstocking, overgrazing and long-term irreversible desertification of extremely fragile dry-savanna and semi-desert ecosystems;

o uncontrolled firewood gathering in savanna forest and bush savanna areas, thus encroaching on the protective tree and bush vegetation of forested, grazing and cultivated areas which have hitherto featured traditional mixed farming;

o the overmanuring, poisoning, salination, alkalinization, compaction and destabilization of soils or groundwater through the excessive use of mineral fertilizers and machines, excessive plant protection and inappropriate irrigation, thus impairing soil fertility, the health of flora and fauna and human well-being in the widest sense.

The underlying causes that link all these agro-ecological problems usually extend beyond the boundaries of individual agricultural zones. They extend into aquatic ecosystems, such as streams, rivers, lakes and coastal waters when, for example, residues of mineral fertilizers and plant protection agents (nitrates, heavy metals or toxins) enter ecosystems with surface or groundwater. The links also extend beyond the terrestrial boundaries into the atmosphere, such as when large-scale logging or slash-and-burn operations disturb the vital oxygen and carbon cycle and thus change the regional climate. Finally, the effects also extend beyond the boundaries of an economic region when, for example, the urgent need for foreign exchange to overcome balance-of-payments difficulties can be met only by resorting to monocultures of export crops, which then have to be maintained to the detriment of appropriate, balanced, self-sufficient agriculture.

It is generally agreed that the complexity of environmental problems in agriculture has increased and that the adverse trends in most harmful factors have accelerated. According to the latest data published by the United Nations Environment Programme (UNEP), of the 1.8 billion hectares (b ha) of pasture and dry forest areas in the developing countries, over three quarters (about 1.4b ha) has been degraded by overgrazing, overuse, deforestation, desertification or soil erosion in some form. The closed initial stock of some 1.2b ha (1980) of tropical forests has already been reduced by 28 per cent (339m ha), exposing their soils to erosion or other forms of degradation. Over 50 per cent (some 300m ha) of the 574 m ha of rainfed cultivated land has been degraded in some way by inappropriate land-use, with soil erosion affecting most of this area. According to the United Nations Food and Agriculture Organization (FAO), about half of the 27m ha of irrigated farmland is threatened by such insidious forms of degradation as salination, alkalinization or waterlogging.

The danger of progressive desertification is greatest in Africa because of the large proportion of land in semi-arid climatic zones. The 64 per cent of land at greatest risk is concentrated in the Sudan–Sahel belt, the Horn of Africa, south-west Africa from Zimbabwe to Namibia and large areas of the Republic of South Africa. The speed at which land degradation spreads is increasing throughout the world. Since the early 1980s it has been estimated at about 20m ha a year; 11m ha of this is due to deforestation and slash-and-burn, 6m ha to desertification and 3m ha to water erosion. The annual losses are estimated at about 26b tonnes of topsoil, the most valuable part of the soil resource in terms of land-use. Annual soil erosion per hectare of farmland fluctuates between 20 and 400 tonnes, depending on geographical location and the purpose for which the land is used. According to data published by the UN Fund for Population Activities (UNFPA) and the FAO, over 100 million people throughout the world are now threatened by serious forms of desertification. In Africa as a whole some 25m ha of land has been rendered unusable for agricultural purposes by desertification in the past 50 years. It is calculated that the deserts currently consume up to 6m ha of land each year. In the opinion of UNEP experts, Africa is a continent in ecological crisis.

Of the more than 20m ha of land lost worldwide every year, some 6m remain permanently unfit for cultivation and thus for agricultural production. Unless steps are taken to protect resources, and preventive measures are introduced within the framework of appropriate farming in the wide sense, the FAO believes that some 544m ha, or about 65 per cent of the 837m ha of rainfed and dry farming areas in Africa, Asia and Latin America is in danger of degradation; i.e. environmentally incompatible land-use has so reduced its yield capacity that food production will decline by about 20 per cent.

Only globally linked causes can explain the worldwide destruction of agricultural and forestry resources; indeed the latest findings indicate interaction of anthropogenic and climatogenic factors to a greater or lesser degree. Rapid population growth; rising land-use pressure; rural poverty and environmentally incompatible economic structures and production methods; excessive commitments to export raw materials; and the distorted social organization of land-use and output have as serious an impact in this context as periodic climatic changes or regional shifts of production to increasingly marginal zones. The most important causal links can be summarized as follows:

○ *The increasing contraction of regional ratios of land to number of people due to exponential population growth.* This results in constantly rising demand for subsistence goods, firewood, timber and water, thus exerting growing pressure on the remaining farmland, water and forest resources, which are declining at regionally differing rates. At the same time, it has proved impossible to create sufficient employment opportunities in the secondary and tertiary sectors to offset land-use pressure.

○ *The general economic structures in many developing countries, geared primarily to urban and industrial development.* The resulting trade, price and subsidization policies may well deter appropriate, environmentally compatible agricultural production. On the one hand, this has led to the cultivation of more and more new land for farming, which depletes resources; on the other hand, the promotion of the integrated, environmentally friendly on-farm economic forces of peasant agriculture and the further development of local varieties, breeds and inputs has been neglected or even prevented. In the final analysis, agricultural resources have been overexploited to finance the urban sector.

○ *Unequal land distribution, particularly in Latin America, southern Africa and parts of south-east Asia.* The perpetuation of traditional agricultural structure and land-tenure systems has prevented farms from adjusting flexibly to a rapidly changing economic environment; this has both encouraged the concentration of land ownership and increased the impact of population pressure and the overuse of marginal agricultural land.

○ *The dependence of agricultural development in many developing countries on immoderate national export strategies.* These usually concentrate on a few agricultural raw materials, such as cotton, groundnuts, tobacco, coffee, timber and beef, with no account taken of the fragility of resources. With agricultural systems that frequently do not include good soil conservation, this has resulted in a continuous decline in soil fertility, which poses an acute threat to food production and has caused widespread rural poverty in many countries.

○ *The fact that development co-operation and other factors have encouraged the adoption of inappropriate modernization and growth strategies and technologies in agricultural development.* The spread of intensive agricultural practice in the first two development decades was accompanied by increasing awareness of the disastrous alienation of people in industrial societies from nature. However local innovative forces in the South were not adequately mobilized to preserve nature and were often actively suppressed.

○ *Weak representation and the absence of scope for the peasant and nomadic population to bring influence to bear on agricultural policies, which might have freed them from the current economic necessity to overuse resources.* There is a general lack of structures to facilitate reform efforts between the macro and farm levels, of safe and appropriate concepts for ecologically sound improvements in production and of support for the necessary social organization at grassroots level.

The causes and diversity of destruction of resources are so complex that it is necessary to consider the most important key variables in this context. They form a socio-economic and socio-ecological framework, and include factors which countries should take seriously if they aim to stabilize resource use. This framework includes:

○ the ratio of land to number of people as the aggregate expression of the remaining scope for farm development before an agricultural region's carrying capacity is exhausted;

○ soil fertility as a complex measure of yield capacity, regenerability, sustainability of land resource yields and appropriate farming *per se*;

○ price-cost structures as the most important economic lever for the adjustment of farms and the introduction of technological innovations, change over time and the development of farm systems which aim to earn sustained rents;

○ the land-tenure system as the constituent framework for the ability of the whole sector to adapt farming methods to changing economic and social structures as one generation succeeds another.

47

Holistic approach to the stabilization of resources

Of all the renewable resources of the biosphere, agricultural land, including its water resources, is among the most important because it is vital to humans and domestic animals. The physical and biological stability and sustained fertility of the soil largely depend both on prevailing soil–climate conditions and on anthropogenic influences. There are strict limits to the degree and form of human interference which soil substrates, ground, soil and surface water, flora, fauna and micro-organisms can tolerate under temperate, let alone under sub-tropical, soil–climate conditions.

The yield capacity of agricultural resources is therefore ultimately a function of pedological and climatic (abiotic) plant-, animal-, and soil-biological (biotic) and socio-economic (anthropogenic) influences. However, their sustainability is guaranteed only by a fourth factor, which has hitherto attracted little interest: yield stability over time. Only where these four factors are in balance is the sustained economic carrying capacity of soils ensured. Only if all four factors are considered seriously and simultaneously will the natural regenerability inherent in renewable resources be maintained.

This self-renewing force of natural resources has an intrinsic value over and above the purely physical and biochemical processes which soil undergoes. The preservation or re-storation of short and long nutrient cycles (biotic dimension), the preservation or restoration of the natural regenerability of ecosystems and of soil fertility (abiotic, biotic and time dimension) and, therefore, the preservation or restoration of the carrying and yield capacity of land-use systems (anthropogenic dimension) are preconditions for sustainable agricultural development. They are the beginning and the end of all further deliberations on the stabilization of agricultural resources and on the maintenance of biodiversity of genetic potential. These aspects become all the more necessary economically, as land becomes more limited during economic development.

Stabilizing agricultural resources and maintaining soil fertility, biodiversity and genetic potential are thus the main long-term tasks of the societies and governments of all countries. They are rightly given top priority in global environmental protection programmes. However, these tasks will require enormous productive forces; pooling of widely scattered know-how; innovative intelligence and, above all, a

48

new awareness in dealings with natural resources. It is also important to recognize that the protection of agricultural resources is not an end in itself; its goal must always be the security of supplies of food and agricultural raw materials.

With population growth obstinately high, demand for food increasing in Third World countries and the terms of trade between agricultural exports and imported inputs continuing to decline, farms are still under pressure to increase production. Farmers wanting to safeguard their incomes in this situation are bound to become mass producers and quantity adjusters. Increasing productivity, achieving economies of scale and reducing unit costs are therefore the principal maxims of farm management. They are also likely to dictate farming methods in the forseeable future, as if the aim were to make increasingly intensive use of the supposedly unlimited productivity of nature solely with the aid of technical advances and economic calculations. Although this is not yet true of subsistence farming, the time is not far off when it too will come under enormous pressure to raise productivity as land becomes increasingly scarce.

This excessive concentration on the use of nature for economic purposes must, however, be abandoned in view of the progressive degradation and destruction of resources that have resulted. The compulsion to adjust land-use systems primarily to the prevailing economic climate must now give way to the urgent necessity to adjust them to the natural and economic and thus complex economic–ecological parameters. A more appropriate definition of a sustainable agricultural unit in a market economy is, therefore, as follows: a farming system is a complex of mutually dependent subsystems, such as the soil–climate–water system, the plant system, the livestock system, the technology system, the economic system, the social system and the system of institutional organizations. These must be used in combination to achieve an economic surplus product whilst preserving soil fertility. The priority task for agriculture's threatened resource base now becomes stabilization in conjunction with sustainable security of production. In principle, this could be achieved in three ways:

○ through the development of environmental and agricultural policies that encourage the economic adjustment of farm development to meet minimum ecological requirements;

49

- from within the organization of land-use or farming systems, by taking greater account of the impact of local conditions on agriculture, and by integrating farming and environmental needs, the eventual aim being the creation of environmentally compatible integrated land-use systems; and
- from outside, by offering farms, groups of farms or local communities incentives to stabilize and protect resources.

The different agro-climatic and economic conditions in the Northern industrialized countries and the developing countries call for different approaches to development in each case. The differences in relative economic importance of the agricultural sector in land-tenure systems and in factor ratios necessitate different intensities of land-use. However, the preservation of soil fertility, stabilization of resources in general and the security of raw material supply, particularly in the South, remain paramount objectives.

Maintenance of soil fertility as a task for development policy

Measures taken to maintain soil fertility in the context of appropriate farm development and integrated land-use systems implicitly include a resource-stabilizing objective. They could therefore well become a universal requirement to be met by changes in agricultural development policy in the coming decades. Their importance for development policy cannot be over-rated on economic, ecological and socio-cultural grounds. The conceptual spread of appropriate farm development is currently due in very large measure to the farming systems development approach. It is therefore essential for agricultural policy in regions where resources are under threat to take account of the principles of appropriate farming. This should have clear implications for agricultural research, teaching, training and extension services. Far more importance must be attached to providing support in these areas through development co-operation than has been the case in the past. It is commonly held that investment in this sector will have favourable long-term effects with little likelihood of mistakes being made.

Soil fertility, which has been neglected in agricultural development, is a vital, easily quantifiable and thus assessable natural asset. It is made up of the following components:

50

○ the humus content of the soil, and the formation of clayhumus complexes and highly cohesive soil colloids, reflecting the soil's ability to withstand erosion;
○ the depth and texture of the A, B and C horizons, reflecting the rooting potential of the soil;
○ pore volume, reflecting good soil ventilation, humification, mineralization and nutrient release;
○ sorption capacity, reflecting the soil's ability to bind nutrients, which is in turn heavily dependent on the pH value;
○ cation exchange capacity, reflecting the soil's ability to release nutrients from its particles; and
○ water retention capacity.

The decline in soil fertility caused by inappropriate farming is quantifiable; the resulting damage to the individual farm and the economy as a whole can be assessed in terms of either the loss of benefit to the farm or the decrease in gross agricultural product. The crucial factor in this assessment is the price which a producer or society is prepared to pay for the maintenance of soil fertility, or the cost which a society is prepared to accept on a long-term basis for land degradation and all that it entails.

There are basically three ways in which to influence the improvement of soil fertility for sustainable land development, by ensuring that general conditions are favourable:

○ giving preference to the promotion of land-use systems which already make a relatively significant contribution to the improvement of soil fertility because of their system-integrating forces, e.g. perennial crop, mixed farming, environmentally compatible irrigation, use of fallow, agro-forestry and silvipastural systems;
○ encouraging the production of crops which increase soil fertility, e.g. legumes, mixed crops, environmentally compatible local varieties, field vegetables and perennial crops, and the keeping of local breeds of animals;
○ promoting ecological farming methods and soil and land improvement measures to halt resource degradation as a whole.

Most of these areas concern, in effect, preventive resource protection measures, which can be taken as part of the development of appropriate farming systems. They thus

51

conform to the concept of the sustainable development of agriculture. The economic principle can be complemented by the ecological principle in agricultural economics, primarily by strengthening what are known as the integrating forces within farming systems. These forces are released where: there is a healthy sequence of crops; where manure and fodder are kept in balance by combining livestock production and ley farming; where the risk is spread by having a variety of forms of cultivation and production; where a seasonal balance is achieved in the work to be performed; and where a high level of self-sufficiency is attained. An added factor in recent years has been integrated pest management, which, like all other integrating forces, is intra-farm in nature and an important factor in efforts to achieve mixed agricultural production and thus to maintain soil fertility.

Distribution of the cost of resource stabilization

Among the internal and external costs which arise in any form of agriculture which depletes soil fertility are the following: the erosion, salination and compaction of soil and the leaching of nutrients, which result in insidious losses of soil fertility and of yields; contamination of groundwater; the eutrophication of water; land clearance; and the loss of species, which imposes a burden on the economy. On the basis of a re-definition of the value of the agricultural resource known as the 'soil–climate–biosphere complex' — a finite, degradable and also positional (micro- and macroeconomic) asset — an additional item of expenditure should now be introduced into the overall calculation of agricultural production to cover the cost of stabilizing resources as land becomes increasingly scarce and the drive to use more land increases.

On-farm environmental costs could be largely avoided by taking the precaution of adapting farms to suit ecology and geographical location, but where they occur they should be shared by farms and society. External environmental costs could be regulated by imposing appropriate laws, although they would be difficult to monitor. In developing countries, on-farm environmental costs play a far more significant role than external costs, while in the economies of the industrialized countries both categories of costs have grown steadily in importance. In general, the destruction of the environment can be curbed only if external costs can be turned into internal costs for those who cause environmental damage.

The internal ecological costs caused by a form of agriculture which destroys resources can and should be calculated as resource depreciations, from which soil-stabilizing measures should then be financed. These measures could be paid for either by the farm, in which case yields and incomes would be reduced in the short term, or by society as a whole if the agricultural sector is unable to meet the costs on its own in the long term. To reduce ecological costs to the economy, statutory conditions or taxation policies would be needed. On the other hand, to promote a form of agriculture that stabilizes resources and to reward the achievement of added ecological value, incentive policies should be introduced. Where the form of agriculture is appropriate to the local environment and the farmer, the guardian of a region's natural assets, deliberately employs more labour and uses environmentally compatible inputs and means of production or invests additional capital in his land, society should be prepared to pay a 'price'. This can be achieved with the familiar instrument of a price, subsidization or tax policy, which, defined in non-sectoral and positive terms, should be seen as better than a policy of simply preserving natural assets.

Producers and consumers should certainly share the cost of stabilizing resources, otherwise no one will feel responsible, and production and consumption will continue to destroy resources literally at nature's expense. The fairest possible distribution of these costs is likely to be the central issue in environmentally oriented agricultural development policies in the future. At all events, depreciations due to ecological factors should be deducted from the gross agricultural product in the national accounts (to give, for example, a net agricultural product adjusted to allow for the environment). Further research work is needed to indicate how the development of production and the protection of resources at different locations can be ensured in the long term with market economy problem-solving approaches.

Because of the ecological handicap, primary agricultural production in particular requires more resource-preserving inputs under extreme soil-climate conditions than production in moderate latitudes if preventive or curative measures are to be taken to protect resources. Such protection normally entails an increase in the labour and/or capital used by production

53

systems. The resource-preserving inputs must be either invested in the soil, through such improvements as drainage, terracing and other land reclamation measures, or linked directly to production through the adoption of such agronomic and grazing management methods as erosion-inhibiting contour ploughing, mulching and the improvement of the soil's water-retention capacity, ecological grazing management (holistic resource management) or mixed farming.

This can take the form of more labour per unit area (labour-intensive), more land per farm (land-extensive) or the substitution of capital for labour (capital-intensive) or of labour for capital (labour-intensive). How the cost of the additional human labour employed on erosion prevention, humus replacement and skilled herd management, and of the on-farm and communal organization of a land-use system that conserves resources should be distributed and who should bear this cost are becoming crucial questions. They also show, however, that it is quite possible for ecological farming, ecological grazing management and even ecological agro-forestry to be so organized that they have an effect on employment. This in turn is very important in development policy terms.

Required necessary agricultural policy developments

As a general rule the developing countries' agricultural policies should seek to ensure sustainable food security and supplies of agricultural raw materials. If a form of agriculture that conserves resources is to be economically attractive, the political, economic and social environment needs to be changed. Such a system should, on the one hand, promote the development of integrated land-use systems appropriate to the given location, i.e. preventive resource protection from within, and, on the other hand, support on-farm and communal curative resource protection measures. The existing domestic instruments of agricultural market price and structural policies and of promoting technical extension services and innovations should therefore take due account of the additional environmental component needed to stabilize overworked and impaired resources.

This could be achieved with a policy of preserving natural assets that combined with the macroeconomic task of ensuring food security and supplies of agricultural raw materials with the task of stabilizing resources. This would require a combination of a policy for the adjustment of agricultural structures and an incomes and prices policy in which a number of priorities is juxtaposed. This policy would thus constantly have to contend with the conflicting objectives of security of food and agricultural raw materials supplies and the protection of resources.

The aim of the policy should be to reward agriculture not only for its economic but also for its ecological added value. Where it is no longer capable of creating the latter by its own efforts, it should be supported by society as a whole. Agriculture which is appropriate to the given location is ecologically and economically more stable because it is more closely linked to the land, operates with greater biological intensity and achieves yields and profits primarily from natural productive forces. It therefore merits the backing of price and subsidization policies more than a form of agriculture that is geared to high yields, is dependent on external inputs and depletes resources. Where a price policy is orientated towards the long-term preservation of the environment, subsidies for the protection of resources are unlikely to be needed, since agricultural economics and ecology will then be in balance.

In principle it should not be the purpose of an agricultural promotion policy to subsidize agriculture which is not viable in the long term and where subsidies actually help to destroy any ecological balance that still exists. An ecologically orientated development policy should therefore divert the widespread input subsidies currently granted with the sole purpose of increasing efficiency and yields at the expense of the environment to government promotion of the sustainable farming systems. Further research should be carried out into the structural, price and trade policy approaches to development that are becoming important in this context, with the following aims in mind:

○ the stabilization of agricultural resources should come primarily from within the farm's own development;
○ this calls for favourable background conditions, guaranteeing not only a reduction in the subsidization of environmentally incompatible inputs but also prices for agricultural produce that cover the cost of stabilization;
○ the cost of any additional investment in land-improvement

55

measures that stabilize resources must be borne by society as a whole;
o the cost of stabilization should be shared fairly between producers (through reduction of subsidies) and consumers (through environmentally appropriate prices);
o the difference in the fragility of resources, as found between the tropical and the temperate latitudes, will give rise to comparative economic advantages as a function of differing stabilization costs; these costs should be taken into account when international trade and food security policies are formulated.

Implications for development co-operation

National and international agricultural policies have key tasks to perform in the attempt to achieve security of world food supplies, in a locally appropriate manner and on the basis of local efforts. The governments of many countries in the South and the North have so far under-rated these tasks. They have yet to tackle them adequately and sometimes feel that on their own they are not, or no longer, able to formulate agricultural development policies which are environmentally and locally compatible. This would require both international agreements among the main agricultural trading nations and support for the protection of natural resources, particularly from the industrialized countries and the international organizations.

While production and tariff agreements attuned to the local ecology should be placed on the agenda for future GATT negotiations, support for the agricultural policies of developing countries seeking to make agriculture environmentally compatible is primarily a matter for development co-operation. In countries where the destruction of resources is widespread it seems clear that development co-operation will be able to play its role effectively only if the measures extend beyond the agricultural sector. Population, agricultural reform, training and extension policy measures should be given as much thought in this context as those taken under agricultural trade policies. In the preventive and curative stabilization of agricultural resources the following aspects are emerging as the main issues for development co-operation:

o Support for international and national efforts to curb exponential population growth in regions where the socio-economic and ecological situation is critical. These policies should be in the shape of family planning and birth control schemes and programmes for the improvement of the social position of women and the development of social security systems.

o Support for the linking of agriculture with environmentally compatible land-use plans designed to ensure security of supplies of food and agricultural raw materials. This should be geared to carrying capacity criteria, yet needs to be developed further, to include such issues as population-supporting capacities and agricultural safe-carrying capacities, and to take account of different intensity and technology levels.

o Support for the development and introduction of economic, organizational, fiscal and institutional preconditions for the application of ecological standards in agriculture, with the object of achieving sustainable and socially balanced land-use. This will become increasingly difficult as government controls steadily weaken in the developing countries.

o Support for the agro-ecological zoning of national agricultural resources, and also with land reform and regional migration and settlement programmes to ease the pressure on overused natural areas. Where unavoidable, the spread of agriculture into pristine wilderness areas should be accompanied by measures designed to afford effective protection to key natural areas. Large-scale transmigration programmes should, however, be eschewed.

o Support for further development of all forms of appropriate farming and integrated land-use systems and their wider application. Instruments for this include agricultural price policy, extension services and the promotion of innovations.

o Support for research, training and extension services relating to the ecologically compatible economic use of agricultural resources. How stabilization costs should be distributed, how agricultural products produced by eco-logically appropriate methods should be priced and whether markets and prices should be regulated are questions that need to be considered carefully and are likely to require international agreements in the long term. Further research on these aspects is needed.

o Support and protection for the marketing and sale of organic-ally grown agricultural products in world markets. These should eventually displace the contaminated agricultural products that result from the use of chemicals and non-sustainable mass-production methods.

o Financial and technical support for the reclamation of degraded land, with soil improvement measures, on the basis of helping people to help themselves and of local and national resource protection programmes.

o Support for international monitoring under UN supervision (UNEP, UNESCO) of endangered areas which do not, or no longer, appear suitable for agricultural use. Their conversion to biosphere reserves should be on the basis of international co-operation.

o Conceptual and instrumental support for all measures to stabilize agricultural resources taken within the framework of sectoral and inter-sectoral approaches to rural development. This would be most commensurate with both the integrating nature of appropriate agriculture and the farming system approach in agricultural development.

Socially, economically and ecologically balanced rural regional development policies must emerge and must prove themselves in practice. This will not happen overnight; it is more likely to take the life-span of a generation. What is important is that we must make a start today and prepare ourselves for a lengthy learning process. This process begins with the realization that there is an indissoluble link between economics and ecology, and that the next generation must not pay the price for the present exploitation of resources.

PART 2: *Latin America and the Caribbean*

Agrarian Reform in Brazil

LINO DE DAVID

The modernization of the Brazilian economy and the transformation of agriculture began soon after World War Two. It took place against a background of a world divided into two blocks, the consolidation of United States power over the Western block, and the consequent expansion of international capitalism to developing countries through multilateral enterprises.

Agriculture went through the greatest transformation of any economic sector. The changes were characterized by the intensive use of agrochemicals, including fertilizers; diversification into different types of crops and livestock (with minor incidence of blight and disease); the intensive use of family labour; and by monoculture in land-use. The 1940s witnessed the setting up of the Rockefeller Foundation and the first beginnings of a Green Revolution. Pilot projects began in strategically selected countries, especially in Brazil, Mexico, the Philippines and the United States.

The Green Revolution had two different stages: the first, an experimental stage, ran from 1943 to 1965. This was followed in the years after 1965 by the great expansion stage. As the revolution advanced, so governments of developing countries plunged into external debt in order to purchase imports. In Brazil, the end of the first stage coincided with the military coup of 1964. The coup linked the interests of powerful multinational groups and big Brazilian firms who, under the protection of military governments, could pursue modernization policies which damaged the interests of most Brazilian people. The policies were opposed by numerous movements who were against their anti-democratic nature.

59

Land rights

The tragic history of the fight of Brazilian rural workers to obtain land as a guarantee for life and freedom, started in the seventeenth century and has passed through different revolts, such as the 'Revolta de Canudos', led by Antonio Conselheiro (1891–5), the 'Pelados war' in the Santa Catarina and Parara (1910–16) up to the 'Ligas Componesas' of the 1960s.

In the 1970s, the fight for land was again taken up by the rural workers, so ending the inactive period imposed by the dictatorship. People claimed land, wages, democracy and dignity; they denounced the atrocities committed in the name of modernization and development — persecution, torture and death — as well as the social and economic consequences of the Green Revolution.

But these movements did more than just denounce and resist current policies. Soon they accumulated forces to build their own perspective of development; this included widespread agrarian reforms, a new technological basis, with a just division of land and property, against a background of diverse farming methods and organization. The cause was just and had popular support, but it had little impact on policymakers. The modernization of agriculture continued.

Today, the imposition of a pattern of development disconnected from socio-environmental concerns has driven Brazil into a corner, with starvation almost nationwide and with natural resources devastated. Agricultural development based on the technological pattern of the Green Revolution, mostly through the intensive use of agrochemicals, hybrid seeds and mechanization, has led to the loss of natural resources such as soil, water, fauna and flora.

Soil degradation is currently so serious that it has led to stagnating yields of soybeans, one of Brazil's major export crops. In the state of Rio Grande do Sul, for example, soybean yields averaged 1391kg/ha in the 1972–80 period, and 1396kg/ha in the 1981–8 period. The yield barely increased despite the widespread use of chemical fertilizers. Moreover, 20.1 tonnes of soil a hectare are washed away every year from the area planted with soybean in the state. National soil loss could be as high as 1 billion tonnes a year. As a result, river and lake silting is increasingly serious, lowering the amount of water availability on the surface and increasing the risk of floods. As the problem has extended to dams, the state hydroelectric power capacity has declined between 30–40 per cent.

Brazilian forests, which used to cover 1.3m square kilo-metres, have now plummeted by 152,702 square km, only 12 per cent as much. And, in several states, the deforestation is even worse. According to The Brazilian Farming Research Company (EMBRAPA) deforestation in the Amazonian region in the past 20 years has destroyed 40m ha of forest. Of this total, 30mn ha was cut down or burnt by big landowners who switched to cattle-raising. In Rio Grande do Sul, the landscape of a large area was totally changed over a few years when a huge subtropical forest was cut down, giving way to wide-open unsheltered areas. Although 10.7m ha have already been deforested, this amounts to less than 1 per cent of the original forest.

In the Amazonian region, big projects such as Transamazonia, Carajas, Balbina and Tucurui have introduced roads, mining, reservoirs and colonization, and caused serious damage to native Indians and to natural resources. Inhabitants living near dam sites have been evicted from their homes, while public debt has mounted owing to sky-high financial costs and the insignificant return on these projects, often only 0.02 per cent.

Deforestation, besides causing soil erosion and river silting, has led to the almost complete destruction of fauna. With no shelter and food, many animal species have become extinct. The most resistant ones have created some defence mechanisms. However, after the chemical offensive, not even these mechanisms were able to resist, and ecological in-balance has worsened.

The intensive, indiscriminate use of chemicals, several of which are prohibited in the countries where they are made, aggravates the problem of water pollution, causing fish morta-lity and contaminating both people and animals, while the residual effects remain in soils for more than 20 years. The widespread utilization of chemical fertilizers (especially the nitrogen-based ones) also hampers soil and water preservation.

Brazil's National Agricultural Credit System provides farmers with interest-free loans to purchase fertilizers, seeds, equipment etc., but its rules are geared to help farmers who grow export crops and who wish to purchase chemical inputs. With such 'support' measures, the consumption of chemicals rose 42 per cent in the period from 1964 to 1979, but food yields increased by less than 5 per cent. The result was that in 1986 the average amount of poison sprayed per Brazilian

61

inhabitant was 3.8kg, seven times higher than the acceptable world average of half a kilogram of poison per inhabitant a year. Meanwhile, the cost of agrochemical inputs caused a 60 per cent increase in food prices. Research in Campinas Sao Paulo (in 1985), involving basic food products such as milk and dairy products, vegetable oils, rice and black beans, showed that 59 per cent of the samples were contaminated with agrochemical residues at higher-than-permitted levels.

Besides causing severe problems to natural resources, the current development pattern of Brazil has impoverished the population. According to the Brazilian Institute of Geography and Statistics (FIBGE), 89.1 per cent of farms with less than 100ha occupy 20.1 per cent of the land. Moreover, the data show that land concentration is rising. In southern Brazil the pattern of land ownership is quite similar, 86.49 per cent of farmers occupy 25.48 per cent of agricultural areas.

Despite the land concentration and the lack of encouragement to small farmers, most basic foodstuffs are produced on smaller farms, as shown in Table 1.

Table 1   Percentage of production carried out on farms

Product	up to 100ha	100–1000ha	over 1000ha
Grapes	95.5	4.1	0.4
Pigs	85.0	13.2	1.7
Poultry	82.2	15.3	2.5
Beans	78.6	18.7	2.7
Potato	75.2	21.8	3.0
Corn	68.2	25.7	6.0
Milk	43.8	44.5	7.6
Soybeans	49.2	40.9	12.9
Cattle	30.7	29.0	30.8

Source: FIBGE

On the one hand, 46 per cent of the country's land is controlled by 1 per cent of owners; on the other, 12 million agricultural workers own no land. In the 1970–90 period, nearly 40 million people were forced to leave the rural areas and increase the poverty of urban centres. The metropolis, as a product of this development pattern, degrades human beings. The benefits of urbanization are completely out of the reach of most people, the big cities exclude them from the right to a dwelling, education, nourishment, health, labour and leisure. Stress, violence, prostitution and death is the lot of many people who drift to the cities.

While the government promises agrarian reform, but fails to provide it, Brazil's problems keep on rising. Between 1985 and 1989, 561 agricultural workers were murdered. Of 1630 murders in the country from 1964 to 1991, only 24 cases went to court, with just 12 people pronounced guilty. Despite technological progress, the level of rural poverty is increasing, not just in Brazil but across Latin America. In countries such as Bolivia, Colombia, Ecuador, Haiti, Honduras, Paraguay, Peru and Venezuela, up to 50 per cent of the rural population lives in absolute poverty; 60 per cent of the deaths of children under 4 years old are induced by undernourishment.

Wealth concentration in Brazil, which was already highly skewed, became even more pronounced in the 1985–9 period, when the national income shared by the poorest 60 per cent dropped from 20 to 18 per cent, while the richest 20 per cent moved from 61 to 63 per cent.

Government policies have encouraged 'modern agriculture', earmarked credit for certain products (soybeans, sugarcane, cotton, cocoa etc.) and opened up new land for cropping and cattle-raising. Loans have gone mostly to big farmers, who were considered capable of handling the new technological package. The only small farmers able to benefit were those who could also show themselves able to adopt the new practices. Export crops using modern technology, such as sugar and soybean, have increasingly expanded their acreage. The acreage under sugarcane rose as part of the fallacious Pro-Alcool project, which aimed to reach self-sufficiency in fuels.

Conversely, the acreages under food crops for domestic purposes, such as black beans and corn, were kept at steady levels.

Table 2 Planted areas in the state of Sao Paulo (in ha)

	1973–4	1978–9
Sugar	790,000	1,163,000
Soybeans	335,000	535,800
Coffee	800,000	1,014,700
Oranges	378,000	516,400
Rice	464,700	300,400
Beans	289,600	351,500
Peanuts	209,700	203,400
Corn	1,290,000	1,054,500

Source: Agricultural Economic Institute (Sao Paulo)

63

In a 5-year period, sugarcane and soybean areas rose 47.32 and 59.34 per cent respectively. Meanwhile, areas under crops for local consumption either rose too little or stagnated, as in the case of corn, whose area dropped by 18.26 per cent. To take a longer time-frame, between 1965 and 1985, the area under soybean rose from 384,643ha to 3.64m ha, an almost ten-fold increase, while the area under beans rose only from 145,501ha to 196,682ha.

Brazil is one of the world's most heavily indebted nations. Most of the lending that caused the country's foreign debt in the 1970s was for the building of large dams, hydroelectric power stations, nuclear plants, and for the purchase of weapons. Money was borrowed for development projects that only benefited top national and transnational groups — irrigation projects in northeastern Brazil for example (several of which never really worked), and for agrochemical imports. Many millions of dollars were spent importing agrochemicals, and the chief beneficiaries, to the detriment of farmers, were agro-industrialists and exporters.

Foreign debt has led to a major worry of recent governments — the performance of agreements with the International Monetary Fund. Strict adherence to these agreements has damaged the living conditions of Brazilians. There is a case for the suspension of external debt payments as well as a fundamental revision of the investment politics of international banks and international finance organizations.

Imports of agrochemicals have hampered Brazil's capacity for food self-sufficiency, besides creating technological dependence, not to mention the environmental consequences. The pattern of farming based on agro-chemicals and the encouragement of monoculture is no longer acceptable in terms of social, environmental and economic costs.

The green revolution type of agriculture has gone through a crisis which it seeks to solve through expansion, for example through the development of the new biotechnology. This would, however, again benefit mostly the agro-chemical industry, enabling it to control and patent seeds, fungi and bacteria. It would give the industry complete control of agricultural potential, concentrating even greater power in its hands.

Land reform

Opposing the expansion of capitalist development and Green revolution-type agriculture are agricultural workers' movements, and non-governmental organizations. They are developing new ideas which they mean to make feasible. Their proposals follow principles in which people and nature recover the right to life, principles that are based on the value of people's knowledge, on coherent and widespread farm reform, the generation of new technological patterns, and the building of new social and working relations, based on co-operation and fellowship.

A new development pattern for Brazil necessarily implies changing the country's agrarian structure. This suggests that a broad discussion on agrarian reform is a fundamental step towards solving the serious political, social, economic and environmental troubles that have hit the nation.

The most important facet of the process of the struggle for land in Brazil during the past decade has been the rise of the Landless Rural Labor Movement (MST). Through the mobilization of farm workers owning no or little land, and its articulation with support associations, MST has advanced the cause of farm reform. This progress is not limited to the matter of land alone, but extends to the organization of farm workers who have been allocated land. A serious and responsible programme of farm reform must respect the agrarian movement and count on the participation of the MST and other organizations that support the need for reform.

In other words, farm reform must be democratic and involve popular mobilization; it must encourage and respect settlement rules, and be promoted through the organization that most suits the agrarian workers, that is as close as possible to their needs. Agrarian reform is not possible without land distribution. In order to settle the millions of landless or nearly landless in Brazil, many million hectares of land are required. Today the land is in the hands of big national and multinational companies and traditionally incompetent large landowners.

When agrarian reform, including land redistribution, takes place it will make a major impact on rural peoples and on food output. It will increase food production, meaning there is no need to open up new producing regions and so contributing to environmental preservation; it will also help to persuade people to stay in rural areas, easing pressure on the big cities. Practice has shown that the best size for new land settlements must follow local patterns. It is also important that the quality of soil of the redistributed land is good so that farming and cattle-raising activity are economically feasible.

The success of the agrarian reform is closely related to the availability of credit to re-settle families. The demand for credit is often heavy in the early years of land reform when a new production structure is being established. Credit may be needed for the recovery and preservation of soil, the purchase of animals and equipment, building of barns for crops, stables for livestock, houses for workers etc. Special subsidized credit needs to be created to help the many thousands of under-capitalized farmers who are on the verge of bankruptcy, thus allowing them to invest more in their farms and increase yields and output. Making small farmers a priority area for credit would help bring about the diversification of agriculture in Brazil towards foods for the domestic market, thus increasing the supply of locally available food.

Subsidized credit is also needed to encourage farmers and co-operatives to substitute conventional agriculture for new methods based on agro-ecological principles, which allow growers to take advantage of organic residues and develop alternative production, storage and processing methods. The great challenge is to produce healthy food in sufficient amounts to feed the farmer and his family, and help to ease the hunger of 70 million Brazilians who are now undernourished.

The results of agrarian reform in countries such as Italy, Mexico, Spain and elsewhere show the increase in yields and food output that can be obtained, and the boost to economies this provides. In Brazil, the experience of agrarian reform shows that food production can increase by unprecedented amounts.

Farm reform is therefore a synonym for higher yields and production, and the key to increased economic activity in rural areas. Technical aid has a role to play in helping to develop the type of agrarian reform that is best suited to the conditions of growers and their environment.

Policy

Government and big money interests have made use of farm policy to implement and ensure the pattern of farming they want to see. For Brazil to get a new physiognomy, radical changes are necessary in current farm policy. It is important that small farmers, who make up the great majority, are able to participate in the elaboration, definition and implementation of the new policy.

Incentives are needed to encourage smallholders to integrate animal and vegetable production, and to adopt practices that need neither pesticides nor chemical fertilizers and which preserve natural resources. As farming is subject to losses induced by disasters such as hail, drought, frosts and winds, farm insurance is needed to protect and ensure investments.

It is impossible to think of a new production pattern in Brazil without considering a restructuring of farm research. To begin with, it is necessary to democratize research, making it relevant to the majority of people. As far as possible, laboratory research should be replaced by on-farm investigations, drawing on farmer knowledge and experience, and integrating existing initiatives. Research should abandon agro-chemistry and turn instead towards enquiries into natural renewable resources (agro-energy and agro-ecology), and break the technological dependence of monopolies and Western countries.

Brazilian farmers lack guaranteed markets and prices for their products, and face major uncertainty over which crops they should plant. This results from lack of government policy with regard to buffer stocks. The construction of storage facilities would enable strategic food stocks to be built up and ensure food supplies to people without sharp price variations, even in periods of disasters or crop failure. Moreover, the guarantee of food availability, by means of buffer stocks, would help to control prices to the benefit of consumers.

A new pattern of agriculture is possible only with a radical departure from the practices and concepts of the Green Revolution, which conceives the Earth as a mere physical support, from which goods are pulled out and profit maximized. The new agriculture should focus mainly on the production of healthy food, based on a diverse ecologically based system. A new development pattern for agriculture presupposes animal/vegetable output integration, long-term sustainability, energy balance, and the diversity of animal and vegetable species.

As soil is the basis of farm production, it has to be properly maintained in order to ensure sustainable food production. Therefore, recovery, conservation and treatment have to be dealt with jointly; it is no use recovering soil fertility without investing in conservation and involving farmers in the best ways of maintaining soils. Recovery and conservation of soils, by organic fertilization (organic residues, manure, crop remainders), green fertilization, during both summer and

67

winter, and natural fertilizers (basalt, phosphate), have brought excellent results. These practices allow a constant supply of nutrients to plants and organic material for soil micro-life, plus coverage of the land area. They protect soils from the impact of rain, thus reducing erosion and the silting up of rivers and lakes. The practices by themselves are not enough however. Other preventive steps have to be taken, among them the building of terraces, withholding strips, planting in strips, always respecting river micro-basins.

Agricultural co-operation is expanding in Brazil, mainly among small-scale producers, as a means of resisting adverse forces, and in the search for better results in the food production process. Such co-operation might consist of groups of families organizing themselves in different ways to mobilize for a particular cause, for example, to purchase equipment for the use of the whole community, to partly collectivize production, for transportation, storage and processing. More recently, especially in the settlements, many CPAs (Farming Production Co-operatives) have been formed, and are planning what, how and where food production will be managed by the whole community.

In the CPAs, families normally live together and develop new relationships. Practical experiments are discussed by the members. Technical aid for the groups reaches as many farmers as possible, farming follows agro-ecological principles and attention is paid to raising the quality of the labour force available in the groups.

Organization into groups is important; it means that credit is more easily obtained, which in turn allows higher investments in food production and the adoption of technologies that would be impossible or unfeasible on an individual basis. Storage and processing facilities are more efficient if they are used by several groups. Collective buying and selling allows for price bargaining and makes the best use of transport.

Self-management of groups encourages creativity, and stimulates the search for alternatives, and also for market information, trading opportunities, processing, improved management, and so on. All of which is an educational process.

If the big money behind Brazil's development has brought great advantages to some, it cannot hide deep social wounds, destruction of natural resources and the impoverishment and degradation of the living conditions of most of the population.

We need to build on the historical know-how that has been accumulated by Brazilian people, and offer a new style of development, combining traditional wisdom with scientific know-how for the construction of a just and democratic society.

Colombia: Save the Rainforests — give them away*

PETER BUNYARD

It is impressive to travel the 500 kilometres upsteam from La Pedrera, close to the border with Brazil, to the old penal colony at Araracuara, plum in the middle of the Colombian Amazon. Except for the small clearings of the occasional Indian settlement, the forest on either bank of the 1km-wide Caqueta river is intact. Moreover, flying over the region shows a continuous expanse of forest stretching as far as the eye can see. For once official data appear to be correct in stating that of the 38.5m ha total area of the Colombian Amazon, more than three-quarters is still forested.

I went to Colombia to see for myself the dramatic changes taking place in the Amazon as a result of the government's policy to grant title to the lands of its indigenous inhabitants. One of my first trips was out of Mocoa, in the Upper Putumayo, from where I walked along a winding, precipitous mule track to visit an Indian community living on a small plateau overlooking the Caqueta river as it plunged down the foothills of the Andes to the Amazon plains below.

There, I saw the effects of centuries of exposure to missionaries and more recently of colonists upon the 400-strong community of Quechua-speaking Inga Indians. That was to be a reminder of how destructive Western civilization could be not only of cultures but also of the environment. By the time I arrived, the Indians of the Colombian Amazon had received title from the government to 12m ha, the bulk of the land being in the central region of Colombia's Amazon territory.

This remarkable act was followed half a year later by the granting of title to a further 6m ha, thus putting into the hands of 70,000 Indians a contiguous area of land that was more than three-quarters the size of Great Britain and was the largest

---

* This is an adapted version of an article in the September/October 1990 issue of *New Democrat International*.

government-granted territory to indigenous peoples of any country.

Colombia has therefore led the world in recognizing its obligations to its indigenous peoples. Moreover it has done so in full recognition of its intention that its Amazon rainforest should be preserved, at least the bulk of what was left. Colombia has instituted the granting of title to the Indians of the Amazon without any pressure from outside. There have been no debt-for-nature swaps.

In fact, the Colombian government appears to have fulfilled certain of the requests of the South American Indian Organisation (COICA) that conservation of the Amazon rainforest should be directly linked to recognition of indigenous rights to land.

During my stay in the Amazon I joined up with Indian leaders from more than 20 different communities from the lower Caqueta, Apaporis and Miritiparana rivers, who were attending a special meeting to discuss how the new policies for the Amazon would affect their communities. We stayed, some 50 of us, in a magnificent *moloca* (the communal house) that had been constructed a few months before. During the week of intense discussions it became clear that the Indians accepted fully the role put upon them by government for caring for the rainforest. In return the government has enshrined in law the fundamental rights of the Indians to their own cultures and traditions. In fact, these were to be encouraged, being considered part and parcel of ecosystem conservation.

Intense pressure

The Colombian government's far-reaching policy to protect the remaining rainforest from further destruction has been carried out despite intense pressure to develop its Amazon, from agricultural, mining and logging interests. What makes the Colombian initiative unique is that the land put officially into Indian hands is based on their extensive use of the forest and not on some Western notion of the land being subdivided into small agroforestry plots.

Over the past decade, lawyers, anthropologists, biologists, politicians and Indian leaders have worked together to find a legal formula that would enable the government to grant all the Indian groups of the Amazon inalienable rights to their traditional lands.

Through resurrecting old colonial laws, Colombia has now found a legitimate mechanism that concedes the entitlement of the indigenous communities to land to which they have *a priori* rights and equally grants them considerable autonomous status within Colombian sovereignty. For instance, indigenous communities in Colombia now have full rights to bring up their children according to their own traditional medicines, and to use their own traditional systems of authority in holding their communities together.

And unlike other citizens of Colombia, indigenous peoples are automatically exempt from military service. Colombia has an estimated 450,000 indigenous peoples, approximately double the numbers officially recognized in neighbouring Brazil. Of that total, some 85,000 have yet to receive title to lands.

On the other hand, although the government recognizes that 210,000 Indians in the Andean region of the country have title to lands that go back to colonial times, much of the land is occupied by non-Indian landowners who have settled there over the past two centuries. The government is buying back some of this land for the Indians, but the process is slow, owing to the high cost involved and the vexed problem of persuading long-established landowners to relinquish what they by now consider to be their own lands.

Ironically, the government has been heavily criticized for having negotiated directly with the Amazon communities rather than through the national Indian organization. But had there been bargaining involved in the granting of lands, whatever the rights of the Indians, it is likely that the process of transfer would have become bogged down, especially since the president himself has to commit his government to taking up the issue of indigenous land rights.

As it happens, President Barco gave his blessing to a massive transfer of land rights just at the time when he had called for an out and out war with the cocaine barons of Medellín. Any problems with the transfer of land to the Amazon Indians would probably have led to his shelving the process and maybe would have put back the acquisition of land rights for many years.

72

Key people

Two key people in President Barco's administration are anthropologists. Martin von Hildebrand, who has dedicated much of the past 17 years to working with the peoples of the Colombian Amazon, served as head of indigenous affairs. He has been instrumental in the Indians obtaining legal recognition of their communal property over the 18m ha of *resguardos* (traditional land rights) that is now theirs.

Carlos Castano is head of National Parks for INDERENA, the department of the environment within the government. He has fought with some success to get the national park system extended throughout the Colombian Amazon, as evidenced by the creation over the past four years of the 575,000ha Cahuinari national park, the 1.2m ha Chiribiquete park, the 855,000 ha Nukak park and the 1m ha Puinawai.

At present, as a result of the new national parks in the Amazon, Colombia has some 43 parks covering 10m ha of the country. But Carlos Castano is concerned that the national park system, as it stands, is insufficient to guarantee protection of Colombia's extraordinary species diversity. He has called upon the government to create immediately another 43 parks and ultimately enough to cover 117 new areas. The borders are particularly interesting, not just because of species richness, but also because in most instances they are home to indigenous people who traditionally know no such frontiers.

Two parks at present are bi-national, Los Katios on the Panama border and Tama on the border with Venezuela. Meanwhile an Indian reserve, the Tobar Donosa project, straddles the border between Colombia and Ecuador where it encompasses some half million hectares and is home to 10,000 Awa Indians. One park under discussion for the Amazon would involve territory from Colombia, Peru and Brazil.

Colombia has tackled head on the issue of any conflict between conservation and indigenous rights to territory and to the use of natural resources within a conservation area such as a national park. Where indigenous territory underlies national parks, the communities using the land have full rights to continue their subsistence activities according to traditional uses. They do not, however, have rights to extract natural resources, whether timber, forest products or food for commercial gain. Equally, outside the national park areas, in the remaining Amazon, any commercial enterprise must not be carried out in a way that leads to deforestation or to dwindling natural resources. These are regulations laid down by INDERENA.

Unusual step

Why should the Colombian government have taken the unusual step of stating that the rainforest is best looked after by its indigenous peoples? One of the main reasons is undoubtedly the manifest destruction of the forest since World War Two. Of its total 1.1 million square kilometres, some 0.8 million square kilometres have soils and climate that best support forest. However, today, only 0.5 million square kilometres are forested, some 37.7m ha have been cut down between 1960 and 1984. Deforestation over the past 30 years has oscillated between 660,000 and 880,000 hectares a year. Such a rate, if continued, would see most of Colombia's rainforest vanish within 50 years.

The Colombian Amazon, meanwhile, has been losing forest at the rate of 100,000 ha a year, a rate which, if continued, would deforest the entire region in several hundred years. In Colombia, as in other Latin American countries, most of the forest is destroyed to make way for cattle ranching. Where tropical forest has gone, the land in 93 per cent of the cases has been turned over to cattle ranching.

The carbon dioxide emitted each year from the destruction of Colombia's forests puts per capita emissions on a par with those of an industrialized nation such as Britain. The Indians believe that both the animals and plants are similar to them, maintaining among each other alliances involving exchange and reciprocity. They thus respect the territories and way of life of their fellow creatures. They know how each behaves and they fear the consequences of any abuse. Hence, when living within their traditions, the Indians will only hunt or exploit other beings after asking for permission under the guidance of the shaman who is constantly evaluating the state of the environment.

The coming into contact with white dealers and the savage exploitation by the rubber barons, which led to the decimation of many indigenous communities in the Colombian Amazon at the turn of the century, the establishment of Catholic and Protestant missions, and the need for goods such as steel axes, machetes, outboard motors and the fuel to run them, have all contributed to the eroding of traditions and customs and a loss of respect for the traditional leaders. The priests tried to get the Indians to give up their communal living and forced parents to send their children into the mission schools where the process of acculturation could finally be accomplished.

The issue of converting parts of the Amazon into vast coca plantations with processing laboratories hidden away in the jungle has also raised its ugly head. During the late 1970s some Indian communities of the Colombian Amazon did become embroiled in producing coca for the market, especially for the production of *basuco* (crack).

That phase, like the hunting of skins, has almost passed, particularly as production has increased in Bolivia and Peru. The Indian leaders believe that with the proper recognition of their rights and the re-affirmation of their traditions, the temptation to grow coca for the market will diminish even more. Indeed, the production of coca is, as commentators such as COICA have pointed out, primarily a socio-economic issue with peasants in particular struggling desperately to earn sufficient to survive.

The destruction of the forest, some 250,000 ha in Peru alone, for coca production, as in the Alto Huallaga valley, is linked directly to consumption in the United States and Europe. In fact, US scorched-earth policies, as practised during the 1980s in the marijuana-growing areas on the slopes of the Sierra Neveda de Santa Marta in the north-east of Colombia, have had a devastating effect. The spraying of gylphosphate from helicopters, indiscriminate of its effect on human beings, inevitably led to increased destruction of the forest since those who lost their crop immediately attacked another area in their desperate attempt for economic survival.

Within a matter of 20 years more than half the rich rainforests of the Sierra Nevada have gone up in smoke. Colombians point out the irony of the situation in as much as the marijuana consumers of the United States now obtain most of their supplies from California. Gradually the traditional leaders are regaining their authority, an authority that is now recognized by the state itself. Indeed, in what must be unique in the world, the Colombian government is not only actively encouraging the indigenous communities of the Amazon to return to their traditions and cultures, but has given them the space and authority to do so.

At least in the Amazon, Colombia appears to have got its

values straight, and the word is spreading. Bolivia is now seeking advice from Colombian lawyers on how to create *resguardos* for the Indians of its Amazon region. Who else will follow suit?

Dominican Republic: Slowly does it; learning and teaching about agriculture

MARK FEEDMAN

Development means something far beyond economics, far beyond the increased production of goods and services. It means the development of people; the unfolding of their innate abilities and the liberation of the human spirit. E. F. Schumacher

Usually, in analyses of development problems, the thing that is recognized as the human spirit is left out of the equation. That is why so many development efforts have failed. Until the spirit is taken into the equation, every effort will fail. We try to incorporate the human spirit into our own work.

This chapter has three aims: to ascertain whether biological agriculture works; to look at the human question of how to teach biological agriculture, and how people can be persuaded to take it up for themselves; and to see how they, in turn, can teach these principles to other people.

For many years, our project has been concerned with one fundamental issue. Once the policies of land reform are in order, once money has been appropriated and bureaucracy arranged, who is going to do the work? Is everybody going to leave their offices, go out into the fields and become farmers? All the mechanical elements of land reform can be put into place very quickly, but preparing people to go into the fields and work side by side with their neighbours is another matter. It takes education and training; even worse, re-education, which can take twice as long as the original training.

A rural teaching centre in the Dominican Republic

Our centre is located in the Dominican Republic, on the central mountain range of the island, close to the border with Haiti. It was derelict when we took it over. Within a year we had begun building up crops, constructing a fishpond, and putting in intensive vegetables and a field-crop system. Since then, the farm has developed steadily. The fishponds are now producing both *Tilapia* and carp; pigs and cows are kept in a zero grazing system; coffee pulp, manure and crop waste are all used in compost, and currently 100,000–150,000 pounds of compost are made yearly. Previously eroded and badly degraded land has been brought back into production. The soil has maintained, and increased, its productivity over a ten-year period. All the work on the farm is done with simple hand tools, primarily the spade, hand fork and machete.

The biological farm as a teaching instrument

However, this is only interesting in the context of whose hands are using the tools. The process of working with people is the most interesting and exciting aspect of the project. The philosophy of the programme from the beginning was that we were all learning and we were all teaching.

When my wife and I began the project, we needed to start by learning ourselves. Our teachers were and are the men and the young people who are working with us at the centre. Some of the original people who started with us are now on the staff of the centre, serving as instructors, technical trainers and the area managers of the programme. Their students are the young people from our village. These students are all young women and men who are studying with their uncles, brothers and friends who are now also their teachers, instructors and role models. We are definitely not interested in merely producing farmers; this programme is about local and community leadership, which has to do with human development and self development.

The 25 young women and men currently in the programme have their work divided between 70 per cent practical experience, that is, learning in the field with their teachers, while the rest of their time is spent in academic studies. The first nine years of the project were devoted to developing training techniques. However, this is a human development programme, and although training is one side of the work, to use the training techniques with knowledge and mastery requires academic skills and the ability to understand the concepts that underlay the techniques. The trained people can then go anywhere, to another field, another valley or another part of the country, and begin to teach, using these techniques in different situations. They are doing this already. We are not producing a technique which is a cook-book recipe, but one that requires understanding about exactly what is going on. An educational process has begun.

Taking the process further forward

Taking this educational process one step further forward is tremendously difficult, and is the next step in the project. Progress has to come out of the culture, the background and the particular area. There is no formula that I or my colleagues have found for education. We have to create an education out of a particular situation.

The students begin to teach immediately. The parents of our 25 students are periodically invited into our centre; they tour the centre and have things explained to them by their children. It is very important not to create a gap, perhaps more an abyss, between generations; the children no longer respect the parents while the adults are confused and don't know what is going on. This children–parent teaching is all part of the extension programme, which is carried out very slowly.

Speed is meaningless. A German educator wrote a book called *Planetary Service*, and used the phrase: 'It cannot go slow enough'. With all the speed and all the urgency we feel, we must remember that *it cannot go slowly enough*.

The project in the wider community

The instructors and teachers are now taking these ideas out into the village. For about four years, the life of the village was effectively halted because, just before the last election, the government promised new houses for everyone. Government officials came and knocked all the existing houses down, destroying the organic structure of the community, and eventually, just before the election, they started giving out some of the new houses. This is not an uncommon rhythm throughout the world, I think! Once houses had been allocated, and people knew where their own houses would be, gardens started springing up throughout the community. Other people in the village had been watching the progress of the centre and domestic and commercial gardens came into being.

The whole village is officially a land reform settlement. Recently, a group of young landless farmers requested some land from the government, which was eventually given for a biological farming project. Their work is currently being helped by two Swedish volunteers, along with two former students of the centre who are helping with soil conservation on steep hillsides.

The centre also does a lot of other training. We teach international development workers what needs to be done on the land. Although they come to teach, they often end up learning from the young people of the village as well. Peace Corps volunteers have learnt that there is not much point in teaching nutrition if you don't first know how to grow food. Both Peace Corps volunteers and their Dominican counterparts are taught at our centre.

We also receive visits from many groups. Last year this included a large group from GTZ who were working on a very big project in the Dominican Republic. They sent 90 agronomists and farm leaders to study our techniques. The farm leaders were interested in transferring these ideas to their own area. The agronomists, who get a kick-back from every bag of fertilizer they sell, would not co-operate with the project. These stories don't always have happy endings; at least not immediately.

The centre is working with two other agricultural and rural development schools. The National Evangelical University is co-operating with us to establish a training programme within their Faculty of Rural Development, where the staff want to set up their own centre based on this work. The Agricultural Agronomic Institute, a government institute which is run by a group of Jesuit priests, includes some technical workers from a German aid agency who are co-operating with our project.

In addition, an Agro-Ecological Alliance has just been formed between our centre, which is non-formal and officially unrecognized, the local Agricultural School, about two hours away, and the National Evangelical University. This alliance aims to train young people in the techniques described above, and to establish a training and professionalization programme for agronomists. We call our students 'barefoot agronomists'. The Dominican Church, which is an ecumenical organization

80

and works with perhaps 500 projects all over the country, has a great need for people who can teach these techniques to rural communities, *campesino* groups, women's groups and youth groups.

The alliance will hopefully also create a universal or general curriculum for small-scale biological farming and gardening. Any development agency will be able to turn to the centres which are teaching and training in this common curriculum and be assured of the capabilities of the people they employ. There is now a basis of trust in the techniques that are being fostered and transferred throughout the country.

The centre today

The centre in the Dominican Republic is still growing. Our school house is small and the building is currently being extended to accommodate leadership programmes for young people from other parts of central America and the Caribbean. All building uses local materials. A local agronomist, who graduated from the government institute, is now directing the centre. The school also provides jobs for local people from the village as cooks, cleaners and administrators. Housing has also been built for students and teachers, again using local materials and built in a local style. In addition, a few brick buildings have been erected, using earth bricks made with a Cinva Ram. This is an excellent technique, which we have found needs some experience because all soils differ in their properties, but which can always be made to work with skill and patience. Work also continues on landscaping, and the restoration of earth banks.

Ideas spread out into the community gently and gradually. The road which everyone takes to the village water source passes the centre. The centre has thus become part of their lives. Children of ten or under now have the centre as a lifelong part of their existence.

The project is also about creating a landscape that is beautiful; creating things of beauty that carry us through our lives and bind us to the land with love. Not binding people as economic slaves, but creating a spirit that keeps people on the land because they love the land, because the life in a rural area is a rich life. It is not money that gives us our riches. The wealthiest people cry in their houses as well. It is not money alone that will keep people in rural areas, or lift rural life to a new dimension. It is the values that were there in rural areas before they were destroyed by the processes by which land has been taken away from people.

The wonderful feeling at our centre is largely due to the attitudes and the eagerness of the young students of the village. This comes from within the village, and out of the village, and by the village. We have taken on a task. We do not feel alone; we feel that all of you are with us.

PART 3: *Africa*

Overview

MOHAMED SULIMAN

The elevation of sustainability to a universal goal is perhaps the greatest achievement of the Brundtland report, *Our Common Future*.[1] In today's world the major adversaries of sustainability are excessive consumption of natural resources and excessive pollution of the global ecosystem, our resource base, life-support system and primary source of cultural and spiritual satisfaction. The 'consume more' culture is apparently gaining unmitigated power over the whole world; its spell has subdued eastern Europe, and, travelling with the speed of light, it is invading the humble huts of the world's poor. The spectre of consumerism is haunting the whole world.

The result is that the total flow of resources, from the ecosystem to the economic system, as quantified by population times per capita resource consumption, is conspicuously high in the North, and is growing fast in the South. At the receiving end of this process are the degraded poor and the degraded natural environment. Our common future is being consumed now.

The dilemma is that as we approach the limits of sustainability, the ideology of unlimited, unsustainable economic growth is gaining credibility, while the precarious situation of the poor compels them to seek security in bigger families and in livelihoods that destructively exploit natural resources. Both processes tend to undermine the natural resource base and pollute and exhaust our life-support systems.

The recent United Nations Conference on Environment and Development (UNCED) discussed and adopted measures to protect the environment and achieve sustainable development worldwide, and reached consensus on priorities and agreed concrete measures. But we should avoid too high an

83

expectation of UNCED; although an important step in the right direction, it will be the years beyond 1992 that will witness a continuation of the debate on solving major issues.

The preparatory stages for UNCED saw NGOs of the North preoccupied with issues of the environment, while their counterparts of the South have been concerned with the developmental problems facing their peoples. Yet if we are morally concerned about the needs of coming generations in the South, we should be equally concerned with the needs of the present, living generation in the South. For example, the debate about the future of world fisheries, an important one for Africa, has been overwhelmed by the cacophony of arguments over the moratorium over whaling. Saving whales is important; saving *homo sapiens* is important too.

African economic and ecological systems are today under great stress. The continent is facing a whole spectrum of insecurities resulting from the deterioration or collapse of its political, social, environmental and cultural values and structures. Survival has become a huge burden on body and soul. Africa's economy has degenerated into dependence on a narrow range of primary products; our environment is having to bear the negative consequences of this specialization.

Famine

The most distressing aspect of the African crisis is famine. Silent and not-so-silent famine is a human-made tragedy that could have been and can be averted. But what are the roots of this crisis? For countless centuries, African people have maintained a living balance with their natural environment; they kept an intimate organic relationship with nature, characterized by a higher degree of sensitivity and respect for the workings of natural ecosystems. There was an almost sacred limit to exploitation, and a sense of duty to conservation. Colonialism upset this harmonious relationship by imposing ever-increasing demands on natural resources; it introduced larger-scale agriculture, mainly based on mono-culture, and began massive deforestation and intensive mining.

The African crisis is best reflected in the deterioration in the terms of trade, which fell by about 35 per cent in the 1980s alone. The Brundtland report points out that the global economic system takes more out of Africa than it puts in.

Debts they cannot pay force African countries to over-use their fragile soils, thus turning good land into deserts.

The cumulative effect of land misuse is a major reason for the rampant poverty of rural Africans. Africa's capacity for food self-sufficiency has deteriorated from year to year; food self-sufficiency in cereals now stands at 60 per cent of total needs and is expected to decline to 50 per cent by the year 2000. Per capita food output dropped by 12 per cent from the early 1960s to the early 1980s and the slide continues.

Related to food security are the complex problems of environmental degradation. Desertification has left two-thirds of the continent arid or semi-arid. There is evidence that global climatic change, deforestation and the continuous removal of vegetation at local level are the major causes of eratic rainfall and aridity. The deforestation of Africa is partly due to the expansion of industrial agriculture, mainly monoculture which diminishes biodiversity, often disregards the ecological impact on the land, and harms the environment and the people through intensive expoitation of their natural resources, especially soil and water. In northern Sudan a major cause of land degradation is the uncontrollable expansion of mechanized farming; this increased from a few thousand hectares in the early 1970s to some 6m ha by the mid-1980s.

The cumulative effect of land misuse is a major cause of poverty among rural Africans. The present frantic exploitation of land, forests and water resources, through extensive and intensive use, is rapidly degrading the environment and making large areas of the continent unfit for human habitation. It is certain that increasing aridity cannot be halted unless afforestation is carried out on a large scale.

Structural adjustment

One of the most important causes of environmental degradation in Africa today is the policies imposed by the International Monetary Fund (IMF) and the World Bank under so-called structural adjustment programmes. These favour the growth of export agriculture as the best remedy for Africa's economic ills. In practice, however, the growth of export agriculture often occurs at the expense of local production of food and other commodities essential for people's basic needs. Sudan is an example.

Between 1978 and 1984 the IMF concluded five agreements with Sudan. The export of sorghum was encouraged. Production of sorghum, the staple food of rural people, increased from 1.7 million tons (mt) in the years between 1970 and 1978 to an average of 2.3mt between 1980 and 1987. During the tragic famine years 1982–3 to 1984–5, Sudan, in compliance with IMF policies, exported 621,000mt of sorghum from Port Sudan, while hunger prevailed in refugee camps near the port. The export of sorghum in order to feed the animals in wealthy countries was hailed by foreign economic advisors as the sorghum success story in Sudan.

There was also a land-use conflict in Sudan between cotton and wheat. Prior to Sudan's structural adjustment programmes, wheat self-sufficiency in the country was 48 per cent. After the implementation of the programmes, over the period from 1978 to 1987, this deteriorated to 26 per cent, a direct consequence of the IMF's bias against wheat because it competed within the export crop, cotton. The area devoted to wheat was more than halved to make way for cotton production. But the expected increase in foreign earnings did not materialize. The world price of cotton fell; Sudan lost out on both accounts; both food production and export earnings declined.

When in the mid-1970s the Sudanese government tried to diversify its export base, the World Bank responded by halting its mechanized farming projects in Sudan and funding only the rehabilitation of the Gezira and other cotton irrigation schemes. In the depressed world market for cotton, large quantities of cotton remained unsold.

In late 1991, 9 million Sudanese people were faced with famine, a consequence of failed development policies, civil war and military dictatorship. Famine is replacing other historical stereotypes as the symbol of Africa. This negative image must be challenged with credible alternatives, strategies and visions. No strategy is going to work, however, so long as the burden of huge foreign debts is stifling life. Africa's debt burden has left in its wake devastated economic, social and environmental structures. African countries pay out US $25 billion annually to service their debts. As a consequence, the poor have less food, while soil, water and forests come under greater pressure.

Uruguay Round

The combined toll of structural adjustment programmes and the debt burdens is as devastating and traumatic as the combined consequences of slavery and colonialism. Africa needs peace, democracy, satisfaction of people's basic needs, rehabilitation of natural resources (mainly solid biomass and water) and a return to positive cultural values of solidarity, self-help and equity. But how can Africa develop alternatives against the background of changes to be introduced in the Uruguay Round of the General Agreement on Tariffs and Trade (GATT), and also of climatic change, against which we have barely any defence? The extension of the free trade doctrine could be ever-expanding monopoly powers over ever-extending geographical areas. The Uruguay Round is an attempt to recolonialize the Third World; history is being allowed to repeat itself.

The implication of the new GATT on peoples and their environment could be immense. GATT deliberately ignores all development issues; it does not include the environment as one of its guiding principles, but regulates it to the *exceptions* to the general agreement. People are not even mentioned in the exceptions!

The liberalization of international trade will storm all barriers, national, ecological and cultural; the liberalization of services will accelerate the harmonization of all cultures into one culture, the Western culture. This will pave the way for the total economic and political subordination of the Third World. Liberalization of trade must ultimately lead to the liberation of all production and consumption from all social and environmental constraints.

Sustainable trade has always been and will presumably continue to function as one of the vehicles of enlighten-ment and progress. But the deregulated GATT does not take into consideration the social and environmental impact of its new world economic order. It is foreseeable that the new GATT regimes will intensify the pressure on smallholder farmers to 'grow or go' in favour of agribusiness. The case for smallholder, tree-growing communities can be lost, and with it the most sustainable and people and environment-friendly land-use system in Africa. The relationship between trees, crops and animals has long been recognized by farmers and pastoralists as the very key of survival.

Impending economic and climatic changes will reinforce and aggravate each other. African farmers and pastoralists have led a precarious existence trying to manage the erratic changes of their micro-climate. They have few tricks up their sleeve to combat the massive climate changes that have been experienced in some areas, especially in the Sahel region of West Africa. The Sahel drought was until recently believed to be man-made, the result of huge biomass loss. Since the mid-1980s, however, opinion has swung towards the view that global warming might be the cause. This new hazardous 'waste', caused mainly by industrial societies, now haunts the old dumping grounds. It is too late to prevent global warming from happening. The question is not one of prevention, but of slowing the process and of adjustment. The new GATT will not be conducive to either.

Zimbabwe: Issues arising from the Land Resettlement Programme

JOHN CUSWORTH

A number of issues have arisen from the Zimbabwe Land Resettlement programme. Whilst these are specific to the Zimbabwe programme they are of wider interest and have implications for the rest of sub-Saharan Africa. They are particularly relevant to the debate on land tenure arrangements and farming systems in terms of environmental sustainability, equity and economic development.

The basic justification for the land resettlement programme in Zimbabwe stems from the country's colonial past. This gave rise to the historical imbalance of land ownership between the races inherited at independence in 1980. Approximately one quarter of a million European settlers occupied 40 per cent of the land of the country with over four million Africans occupying another 42 per cent, the remainder being allocated to game reserves and forest areas. The situation was exacerbated, in that 80 per cent of the land suitable for intensive agriculture was occupied by Europeans.[2]

The land issue was of fundamental significance during the independence struggle. It was therefore inevitable that independence would lead to a programme of land redistribution in one form or another. After ten years, and the resettlement of over 50,000 families on more than 3.1m ha of acquired land, the motivation for continuing resettlement remains strong. Given the growing population pressure in the Communal Areas (CAs) of the country it might be considered that pressure to proceed faster with resettlement is, if anything, increasing. Although primarily politically motivated, resettlement has involved wider economic and social development objectives.

Objectives

The broad aims of the resettlement programme were to redress the historical imbalance in access to land between the races, and to create an opportunity for alleviating the

89

economic plight of some of the poorest rural people, whilst maximizing the economic potential of the land. The specific objectives of the programme as set out in the Intensive Resettlement Policies and Procedures Document (April 1985) are as follows:

○ to alleviate population pressure in communal areas (CAs);
○ to extend and improve the base for productive agriculture in the peasant farming sector through individuals and co-operatives;
○ to improve the standard of living of the largest and poorest sector of the population of Zimbabwe;
○ to ameliorate the plight of people who have been adversely affected by the war and to rehabilitate them;
○ to provide, at the lower end of the scale, opportunities for people who have no land and who are without employment and may therefore be classed as destitute;
○ to bring abandoned or under-utilized land into full production as one facet of implementing an equitable policy of land redistribution;
○ to expand or improve the infrastructure and services that are needed to promote the well-being of people and of economic production; and
○ to achieve national stability and progress in a country that has only recently emerged from the turmoil of war.

These were ambitious objectives for a single programme. From the outset, government recognized resettlement as more than simply a means of land redistribution. There were clearly stated welfare and economic aims. These had significant implications for the way resettlement was implemented and for its consequent performance.

Approach

The broad guidelines covering the approach to land resettlement in Zimbabwe were laid down at the Lancaster House talks in London which led directly to the country's independence. Under the agreement the new government would be able to purchase land for resettlement from commercial farmers on a 'willing buyer/willing seller' basis. Payment for the land would be in local currency with prices being negotiated at the appropriate prevailing market rate. If government were to acquire land other than through the willing buyer/seller system then, under the terms of the agreement, it would have been obliged to compensate the landowners in convertible currency.

Given that the new government was to inherit an economy within which there was a need to revitalize many of its sectors using scarce foreign exchange, this clause of the agreement might seem to have been very restrictive. However, land availability through the voluntary channels appears not to have been a significant factor in determining the rate of resettlement, particularly in the early years of the decade.

The day-to-day operational aspects of resettlement are set down in the Policies and Procedures document which was drawn up to assest the various agencies involved with the programme to identify their responsibilities. It is a detailed document which provides guidelines and sets out the methodology for planning the programme on an individual scheme-by-scheme basis. Each scheme was to be planned separately and appraised from the perspective of the settler household and of the national economy.

The 'Intensive' label applied to resettlement indicated the package approach to be adopted to programme imple-mentation, which involved not only the allocation of land to settlers, but the inclusion of infrastructural development and the provision of support services. Resettlement schemes implemented without most of the associated infrastructure and service provision were termed 'accelerated' schemes. These were established to facilitate the rapid resettlement of land pending incorporation under the Intensive programme. The need for rapid resettlement arose as the rate of land purchase was outpaced by the capacity to plan resettlement in the early years of the decade. By 1990 less than 2 per cent of settlers were still accommodated on accelerated schemes.

Productivity

If land resettlement were simply a matter of land transfer between different groups of people with an equal capacity to use it, then the rate of resettlement in Zimbabwe would probably have been much faster. The fact is, however, that the current resettlement programme involves the transfer of land between people with a fundamentally different approach to land–use and with very different resource bases.

The Large Scale Commercial Farm (LSCF) sector consists

of a relatively small number (currently around 4500) of very large holdings (average size 2200ha) deploying relatively capital-intensive, technically modernized farming methods. The sector contributes significantly to exports with practically all coffee, tea and tobacco, and half the cotton crop, being produced on commercial farms. Practically all the sugar cane, soya and wheat crops are produced by the commercial sector, also about half the marketed staple maize crop (although this is declining).

On the other hand the small-scale CA farmers who are intended to be resettled under the programme come from a sector of the agricultural industry that employs relatively little modern technology, produces primarily for family subsistence and concentrates on the production of food crops. This is still generally the case despite recent advances made by CA farmers in the production of crops such as maize, cotton and sunflower. They are now responsible for about half the annual intake of the respective marketing boards.

The significance of the contribution to the national economy of the LSCF sector and the uncertainties over whether small producers could sustain this contribution under resettlement in the short to medium term lay at the heart of the problem faced by government. If it was certain that the output and productivity of the LSCF sector could be maintained under smallholder agriculture, there would be fewer difficulties in determining the rate of future resettlement. The important point is that until productivity on resettlement schemes increases, replacement of the LSCF sector with resettlement will have serious consequences for aggregate agricultural production and for the macro-economy as a whole.

Social development dimension

The current resettlement programme includes a wider dimension than that of the transfer of land from one farming system to another. Whilst it has a productive purpose there is also a very explicit equity objective within resettlement. The programme aims to assist some of the most disadvantaged members of the population. Understandably, war-displaced and affected persons, the landless and returning ex-combatants were singled out for priority for selection for settlement immediately after independence. Resettlement was, therefore, envisaged as a vehicle to address an equity problem and foster social and political stability.

To achieve this latter objective the programme involved the significant allocation of resources towards the provision of social infrastructure' within the programme. The building of schools, clinics and housing for community and co-operative development personnel involved approximately 40 per cent of the development expenditure under the programme. With most settlers living within nucleated villages provided with access to clean water supplies and education and health facilities the programme must be credited with improving the standard of living for a significant number of the poorest people in the country. Whilst this provision of services has benefited some 50,000 households (approximately 400,000 people) its impact has been restricted to resettlement areas. The CA population of over 4 million people has not directly benefited from the service infrastructure established in resettlement areas.

The issue of settler selection is particularly important when considering how to increase productivity. It has been generally acknowledged that many people resettled under the programme are those least likely amongst the CA farming community to be able to make full use of the productive potential of the resources being allocated to them. Had better-resourced and experienced farmers, with a greater capacity to bear the risks of innovation, been selected for resettlement, then current productivity levels would almost certainly have been higher.

More productive settlers, earning higher incomes, might reasonably have been expected to make some contribution to the development and running of the schemes from which they were benefiting. As it is, settlers make no contribution to the upkeep and maintenance of schemes. Increased production, and any contribution to the sustaining of the social infrastructure, would have the effect of creating more resources for government to invest in the CAs where the benefits would be more widely spread.

Additionally, given that most settler households occupied little or no land in their former CAs, resettlement has had little impact on the intensity of CA land-use, as the programme has not resulted in any significant release of land. Neither has resettlement resulted in any absolute decrease in population pressure in the CAs, as the rate of resettlement is outstripped

93

by the rate of population growth. A growth rate of over 3 per cent a year amongst a population of almost 5 million (CSO) provides an increase in population of 150,000 people, equivalent to approximately 19,000 households. This figure puts the scale of the current resettlement programme into perspective, in that the average number of households resettled over the ten-year period is less than 6000. From this it must be concluded that resettlement cannot, in itself, provide a solution to the problems of CAs.

Management and support

One of the main constraints to resettlement has been uncertainty as to whether the production levels and productivity of commercial farming could be sustained under smallholder agriculture. It may be that one way to reduce this uncertainty would be to operate a settler selection procedure which would give more emphasis on selecting qualified farmers to be settlers.

But this by itself would hardly guarantee the desired outcome. It was recognized from the inception of the programme that settlers would need substantial assistance in the early years of settlement if they were to take full advantage of the resources allocated to them. Consequently settlers were provided with a range of productive services such as extension, credit, veterinary services, etc.

For many settlers these services have proved a profound disappointment. While the establishment phase of the programme was conducted relatively smoothly, with the implementation of resettlement fully in line with organizational objectives, it was a different matter over services.

The institutions required to service the productive needs of the settlers, that is, the Co-operative Unions, AFC and the marketing boards, found that resettlement provided them with nothing but extra administrative work for no financial benefit. Institutions such as the AFC and the marketing boards had until 1980 been orientated mainly towards dealing with a relatively small number of primarily commercial farmers. They were quite suddenly required to reorientate their operations towards several hundred thousand clients in communal and resettlement areas. It was not surprising that they experienced difficulties coping with such a dramatic change in organizational culture.

Currently on resettlement each settler is granted three permits: one to de-pasture a certain number of livestock on a communal basis, one to cultivate an arable plot and one to reside on a specific residential plot. These permits are for no specific duration and there is currently no prospect of them being upgraded to leasehold or freehold status.

Whilst there is no substantial evidence that settlers feel particularly insecure under resettlement, there is evidence that settler households are reluctant to cut their ties completely with the CAs from which they came and in particular to release their right to cultivate land there.

This feature of settler attitudes frustrates one of the objectives of resettlement which is to release land in overcrowded CAs and may also act as a disincentive for settlers to invest in the medium- and long-term future of their holdings. Again if the objective is to reduce the uncertainty of settlers being able to match the output and productivity of former commercial farm land it would seem appropriate to provide an added incentive for settlers to better husband the land by providing them with an increased stake in its long-term future.

Land-use

One of the key factors in determining the success of any land reform programme as an economically viable development initiative will be the extent to which land is utilized. For Zimbabwe it has been estimated that although yields on resettlement areas compare favourably with those of CAs, overall output per unit of land area under resettlement is estimated to be less than in the LSCF sector.[3]

Two main factors relate to the use of land: the degree to which land is being utilized at the planned level of intensity, and the appropriate intensity of land-use for the particular area. Ministry of Agriculture survey data in Zimbabwe suggest that there is no substantial under-utilization of arable land in resettlement areas in terms of aggregate land area cultivated.

Evidence to support this conclusion can be drawn from the table showing the percentage of cultivated area in different plot sizes (Table 1). Less than 31 per cent of land under resettlement areas is in plots of less than the planned 3ha whilst some 43 per cent is in plots of over 4ha. This would indicate that it is likely the deficit on cultivated land by the minority of settlers is more than compensated by a substantial number of settlers cultivating considerably more than they were expected to. But this evidence also indicates that there is a skewness in the cultivation practices amongst settlers whereby 51 per cent of settler households cultivate only 31 per cent of the land with 49 per cent cultivating the remaining 69 per cent.

Table 1   Intensity of crop cultivation in resettlement areas

Natural region	Plot size in hectares					
	less than 1	1–2	2–3	3–4	4–5	Over 5
IIA	4	18	15	33	24	6
IIB	6	29	29	14	18	4
III	7	21	23	23	21	6
IV	6	35	24	21	12	3
V	0	44	11	4	30	11
Total	5	25	21	22	21	6

Natural region	Percentage of Cultivated area in Plots of Different Sizes					
	less than 1	1–2	2–3	3–4	4–5	Over 5
IIA	1	9	11	36	33	10
IIB	1	14	24	16	25	19
III	2	10	19	14	31	13
IV	1	22	22	28	20	6
V	0	25	8	2	42	20
Total	1	13	17	25	30	13

Source: Derived from the Monitoring and Evaluation Unit 1988 survey of the 1987–8 season.

If the available evidence indicates that there is no under-utilization of land in resettlement areas in terms of land cultivated, then the issue remains as to whether the planned intensity of arable cultivation under resettlement is appropriate. This issue cannot be treated in isolation to that of livestock stocking rates and draught power. In the planning models the allocation of arable land is uniform across the natural regions with the allocation for grazing being variable. There is an implicit assumption in the planning models for schemes in the higher regions, IIA and IIB, that each household will require a basic cattle holding to provide the animal power to cultivate the arable plot. The grazing land required to de-pasture these cattle is estimated on the basis of the carrying capacity of the land, using technical coefficients appropriate for the production of fatstock under commercial conditions. The number of households to be settled is estimated from the number of draught-power units that can be carried on the land, the limiting resource being grazing capacity.

The intensity of this grazing will, to a great extent, determine the overall intensity of land-use on the schemes. Table 2 shows the distribution of cattle ownership among settler households in 1988. This indicates that more than half of settler households own no cattle, with a further 21 per cent having less than the planned number. At the same time 28 per cent of settler households own herds greater than the planned number and own 78 per cent of the cattle. This highly skewed pattern of cattle ownership has resulted in considerable under-utilization of the grazing resources in some resettled areas.

Table 2   Cattle ownership distribution on sample Model A Schemes

Herd	No. of cattle	1–5	6–10	11–15	16–20	21+
Percentage owners	51	9	12	11	10	7
Percentage cattle	0	5	17	24	30	24

Source: Monitoring and Evaluation Unit Survey 1987–8

Table 3   Percentage use of allocated grazing on resettlement schemes

Province	All As	All Bs	Pre '85 As	Pre '85 Bs
Manicaland	44	13	44	12
Mashonaland Central	75	22	71	12
Mashonaland East	82	41	87	41
Mashonaland West	58	19	53	11
Masvingo	71	39	74	0
Matabele North	68		68	
Matabele South	47	32	47	
Midlands	47	29	49	34
National total	58	24	58	22

This indication of under-utilization of grazing may be overstated through the survey data. Other data collected in

1986 during an evaluation of several well-established schemes showed a similar, if less dramatic, skewness of cattle ownership, but revealed that many settlers ran herds of cattle of much greater numbers than should have been permitted.[4] Some settlers had herds numbering over 70 head. This resulted in schemes being 'overgrazed' on the planning criteria, despite a large proportion of settlers remaining without cattle.

The result of the analysis of land utilization would indicate that the planners were correct in the assumption that draught cattle are essential for settlers to be able adequately to cultivate their arable plots, but after ten years of resettlement and, despite medium-term credit schemes to assist settlers to purchase cattle, a significant proportion of them still do not own any. This group of settlers is, therefore, unable fully to utilize its arable holdings and derives no benefit from their grazing allocation. On the other hand a minority of settler households which own many more cattle than the schemes planned for is able to more than use their arable holding and enjoys considerable benefits from the grazing allocation.

The result appears to be both under- and over-utilization of land in resettlement areas. Furthermore, as schemes mature, the problem of over-utilization will become more intense, not necessarily because poorer settlers acquire more cattle and cultivate more land but because richer settlers keep more than the allocated number of cattle and over-cultivate their plots. The conclusion to be drawn is that there is a certain dynamic nature of land reform programmes that almost always lead to an inequitable distribution of economic benefits amongst the settlers and a varied pattern of land utilization.

Wider implications for Africa

Whilst the issues raised refer to the case study of land resettlement in Zimbabwe a number of them are relevant to the continent as a whole. This is particularly the case in countries where there is a dualism in the agricultural sector in which industrialized high input-high output (primarily export orientated) farming contrasts with relatively low input-low output more traditional smallholder agriculture and where there is an increasing demand for land due to population pressure. The issue of sustained productivity has been identified as central to the argument. As stated above if productivity levels could be sustained regardless of ownership, or mode of production, then the need for land reform could be argued simply on equity grounds. Indeed some studies, including the resettlement evaluation in 1988 of the British government's Overseas Development Administration, support the view that transfer of land from large-scale farming to smallholders actually increases productivity.[5]

However the cost of such reform is beyond the capacity of most governments and there is also a considerable amount of evidence to suggest that the capacity to use land productively varies enormously according to the resource and technology base of the individual smallholder. This problem relates to some of the issues of general interest to agricultural development in Africa, including the following:

Farm size

Most land reform programmes involve the development of a farming system made up of large numbers of similar farms within similar agro-ecological zones. The basis of how big these farms should be will depend on a number of factors but often it is based on a target family income related to the technical coefficients of the local farming system. This seems a logical approach. But if, like Zimbabwe, there is a lack of flexibility to meet the resource base of the household such a system may give rise to the type of land under- and over-utilization described earlier. The end result of this is the failure of the land reform to meet economic objectives and the risk of significant environmental damage.

The issue of what happens to landholdings over time is perhaps more important still. Almost everywhere in Africa the demand for more land for cultivation by smallholders is expanding. This has led to arable cultivation being undertaken in marginal rainfall areas placing people more at risk to the effects of drought, with a simultaneous clearing of forest land for farming and the familiar problems of soil erosion and water resource depletion. Under land reform programmes what happens to the dependants of smallholders as they grow up and seek to earn a living from the land? Landholdings may become fragmented to such an extent that they become incapable even of providing for subsistence. Alternatively if fragmentation is not allowed then the dependants are forced to move away from the land which may in some circumstances (as in the case of the Zimbabwe small-scale commercial farms)

99

end up being under-utilized by an ageing farming population. There are no easy answers to these problems but it is often argued that the answer lies in the nature of the land tenure arrangements.

Land tenure options

There are basically three main options for reforming land tenure arrangements in Africa:[6] 'individual title allocation', 'lease-hold ownership with conditions' and 'communal control of land allocation and land-use'.

The arguments in favour of the first include that of the need to provide an incentive for investment and conservation over the long term. It is also argued that land titles will provide collateral for credit transactions which will stimulate investment. The arguments against are that only title holders will benefit over the short term and that anyone not currently occupying land for any reason would be excluded from doing so forever. Furthermore whilst title deeds may prove valuable as collateral this may encourage reckless borrowing and indebtedness leading title owners no option but to sell out to bigger and richer farmers.

The conditional leasehold option has the attraction of providing the land user with security of tenure. It also limits the potential for fragmentation of landholdings. It may also ensure maximum use of the land. However the drawback to this type of tenure system is that it requires a massive institutional infrastructure to administer it. This is expensive and may ultimately be unworkable in many areas.

The communal control of land allocation and land-use is closer to the systems that operate more generally in most of Africa today. It has the advantage of giving local communities the responsibility for this important function. However this is no easy option. There would be need to develop systems in which the less powerful members of the community, that is, women and the poorest, are able to have a say in the process. Quite how this can be done will depend a great deal on the existing local administration and traditions. A key factor will be the level of awareness within the local community of the collective need to maximize the use of resources in a sustainable manner.

Support services to smallholders

It has been suggested that in the case of land resettlement in Zimbabwe many settlers provided with land for cultivation were unable to make proper use of it due to a lack of both technical knowledge and the means to exploit their new asset. Input supply, marketing and credit systems singularly failed to provide the settlers with the required level of services. This is not a new phenomenon. Virtually all over Africa such support services fail to meet the needs of small farmers. Often this stems from the nature of the farming system itself. The high input — high output type of farming that smallholders are expected to become involved with requires these service arrangements. If they are inadequately provided then the farmers will be considered inefficient and unproductive, thus in turn providing support for those that argue for the transformation of agriculture through increased emphasis on large-scale commercial farming.

The provision of agricultural support services is expensive if organized on a centralized basis. As with the land tenure arrangements some local-level institutional approach might be more appropriate. As many countries in Africa undergo a period of change in economic management such new local-level initiatives must be developed to replace the expensive and often inefficient agro-service infrastructure.

Namibia: Land Reform — who will be the beneficiaries?

MARTIN ADAMS

When Namibia came to independence in 1990, the South West Africa People's Organization (SWAPO) announced its intention to 'transfer some of the land from those with too much of it to the landless majority'. Yet, two years later, more than half the agriculturally usable land in the country is still occupied by some 4200 commercial ranchers, mainly white. The rest provides a home and, in varying degrees, a source of subsistence for about 120,000 black rural households scattered through the communal areas, the former ethnic homelands.

Namibia is a mostly pastoral country; only relatively small areas in the north are suitable for crops and these are frequently stricken by drought. In the south and west, where mean annual rainfall ranges from 50 to 200mm, small stock predominate. In the centre and north, which receive up to 600mm, cattle are more important. Meat is exported to South Africa and the European Community.

The country is finding out that land reform, that is, the redistribution of property or rights in land for the benefit of small farmers and agricultural labourers, is a very difficult change to carry through, especially in a ranching country with strong interests in the preservation of the status quo.

As the government began to consider the practicalities of land reform, it became evident that a great deal of information and consultation was required. Supported by the opposition parties, it decided to conduct a national consultation on the land question culminating in a national conference, held in Windhoek in June 1991. In addition to buying time, the objective was to achieve the greatest possible consensus on the major issues and to make recommendations to government.

The conference, chaired by the prime minister, brought together all the major political organizations and interest groups. Specialists on land-related topics from Namibia, as well as from Zimbabwe and Botswana, were engaged to inform the debate. The views of rural people were expressed

102

in a video based on a survey of opinions on land issues; this was shown at the conference as well as on national TV.

In the run-up to the conference, political groups, representing different ethnic interests, pressed for the restitution of their ancestral lands. These had been lost to the German colonizers at the turn of the century and were now occupied by white-owned freehold farms in the central parts of the country. The passion with which these competing claims were prosecuted threatened to wreck the conference and the fragile process of national reconciliation. However, after three days of debate, the prime minister obtained broad agreement that the restitution of particular areas of land to specific groups was not feasible. This was because ancestral land rights of the various ethnic groups in central and southern Namibia had been superimposed on one another for centuries, if not for millenia, and could not be identified with accuracy.

The conference then moved on to debate the present inequity of land ownership. It recommended to the government, among other things, that foreigners should not be allowed to own farmland, that the land of absentee landlords should be expropriated and that very large farms, and/or ownership of several farms should not be allowed. Other resolutions related to the need to improve the conditions of farmworkers and to resolve land-related problems in the overcrowded communal areas.

For the organizers and the majority of participants, the conference met its immediate objectives, namely to reach a consensus on the land question; this satisfaction was reflected in the media and the mood of the public. Bearing in mind the wide gulf which separated the various factions at the beginning of the conference, the level of agreement reached among participants was remarkable. Can consensus be obtained also on the speed and direction of implementation?

It is still too early to draw conclusions on the implementation of land reform. Land policy has yet to be formulated by the government which is still trying to decide what should be done. Several of the conference resolutions raise fundamental issues with which neighbouring African countries have long been grappling. In retrospect, it can be seen that it was easier for the conference to recommend who should lose land than it was to decide who should gain it. Land for the landless was barely discussed at the conference, despite the vivid pleas in the video.

The view which received the most attention at the conference was that freehold farms should be made available on financially favourable terms to black farmers. The pressure for this reform comes from a number of quarters. It comes from white politicians who are keen to recruit rich and politically influential black farmers into their ranks; from black businessmen and government officials who wish to own farms themselves, and from small farmers in the communal areas who resent the pressure on communal grazing exerted by the large herds and flocks of wealthy stock owners. These small farmers are supported by officials of the Ministry of Agriculture who argue for the transfer of the large herds to commercial farms on environmental grounds. (It should be noted, however, that both environmental and equity arguments for moving larger livestock owners to fenced farms were advanced in Botswana in 1975. The result was negative on both counts).

It was noteworthy that the owners of large herds in the communal areas were not themselves in favour of moving their stock to commercial farms. In the communal areas they enjoy free grazing, water, drought relief and various services without having to pay income tax.

The level of demand for freehold farms will clearly depend on the credit terms available. At current interest rates, 18 per cent a year, demand is expected to remain negligible. The problem will be to target the larger farmers in the communal areas and discourage applications from urban businessmen and officals. Even then a programme of assistance for communal area farmers to buy freehold farms would reach only a fairly small number. It would provide only temporary and partial relief to the crowded communal grazing where the environmental benefits of the programme would be difficult to detect.

Land for the landless

By comparison with the attention given to the relocation of large farmers, the technical and socio-economic problems of providing poor families, landless people, farm labourers and war returnees with land received very little attention at the conference. Namibia is generally dry, and the land is mostly unsuitable for arable cropping.

The proposal to settle peasants from overcrowded communal areas on former commercial ranches raises a number of difficult practical issues. In the light of experience with pastoral settlement schemes elsewhere in Africa, neither the subdivision of ranches into family livestock farms, nor group or co-operative ranching are likely to be viable options.

The cost of settling farmers with small herds and flocks on individual farms, with reasonable standards of social and economic infrastructure, would be very high and the economic return almost certainly negative. In addition to the economic consequences of sub-division there are likely to be far-reaching environmental effects. Small herds, of 50 to 100 head of cattle, are difficult to manage as commercial units. Offtake is much lower (less than half) than from a commercial herd. Initially, herd growth rates would be fast, assisted by the relatively better grazing and by low offtake. But in the narrow confines of the family farm, grazing pressure would be intense and continuous, to the detriment of the herbage and, in some areas, of the soils.

The only viable solution would be to amalgamate several farms, take down internal fences and retain only the minimum number of water points needed for the herds and flocks. If the farms in question are adjacent to an existing communal area, the boundary fence between the two could be removed and the communal range simply extended. This solution tends to be favoured by the majority of stock owners. In their experience, there is no substitute for space to move in. Communal grazing has many virtues, which are often ignored, but the more confined the space, the more destructive it becomes.

However the evidence from neighbouring Botswana is that efforts to improve communal grazing systems by subdividing communal land into separate camps, or by transferring herds from communal land to individual camps on commercial farms, have generally resulted in the accelerating rate of environmental degradation, even if stock numbers are not increased.

Any proposal to extend the communal area is highly controversial. As elsewhere in Africa, official thinking on the development of communal rangeland has for long been characterized by an approach to land-use derived from the commercial farming sector. Agricultural progress, as envisaged by development agencies and African governments, has been confined to transforming traditional stock keepers into commercial farmers and replacing customary forms of communal tenure with individual title. This formula is based on the assumption that traditional techniques of pastoral and livestock management are environmentally destructive and that improvements in husbandry and land management can only be achieved on private farms, a view which has led to widespread landlessness in Africa.

Until now, African politicians and businessmen wishing to acquire ranches with government loans have found allies in mainstream range management theory. However, in the recent words of an authoritative paper, these views 'are fundamentally flawed in their application to certain types of rangeland systems. They require not minor adjustment but a thorough re-examination'.

Agriculture in Namibia employs 60 per cent of the population, but generates only 10 per cent of GDP. With its mineral wealth, Namibia is much less dependent than other post-colonial territories on white farmers and on the export of agricultural commodities to earn foreign exchange. Namibia can probably afford to be more even handed when it comes to land reform, in a ranching country with strong interests in the preservation of the status quo.

PART 4: *Asia*

Land Rights in the Villages of India

VITHAL RAJAN

When discussing land and agriculture, it is important to bear in mind several important issues that are central to the well-being of the rural-dwelling majority in the South:

○ the population explosion
○ the question of debt — the debt burden of the poor mirrors the inability of many countries to pay — and the role of Northern financial institutions in increasing this problem
○ migration, from rural to urban areas, and similar to the movement from the Third World to the North
○ degradation of the environment, and loss of soil, forests and water
○ land rights

The context: villages near Hyderabad in India

This chapter will concentrate on my personal experience, particularly of working at a grass-roots level in a small part of India. Since the 1970s, I have been working in villages near Hyderabad, a major city in the Deccan plateau in the heartland of India's semi-arid tropics. Half the world's human population of the semi-arid tropics lives in about 70,000 villages in this area. The rest are spread through the Sahel region of Africa, parts of Brazil and Mexico, and elsewhere. The per capita availability of food grown in the Deccan region is lower than in the Sahel.

The environment is harsh, and the only reason that there has not been a great famine in recent years is because of the famine relief policy followed by the government. In the nineteenth century, there were a number of severe famines, which killed several tens of millions of people. The relief

107

policy which was then shaped by the British–Indian government was itself based on those instituted in the fourteenth
century by an Indian emperor called Mohammed Bin Tuglak,
popularly known in Indian history as the Mad King. Sometimes
madness has its uses!

The region is environmentally degraded. Forests have been
cleared; loss of topsoil is severe, and in some watersheds
reaches levels of a hundred tonnes per hectare a year;
reservoirs built for power generation and provision of drinking
water are silting up; and the people are poor, illiterate,
malnourished and unemployed.

The Deccan Development Society

My friends and I formed an organization called the Deccan
Development Society to address these issues. The society now
works in 40 of the smaller villages. Forty villages is not much,
less than 1 per cent of the Indian villages, but it still affects a
hundred thousand people.

After many years of effort, we have been successful in
providing the catalyst for the development of a number of
peoples' *sangams*. A *sangam* is, broadly speaking, a local
association; it is a very old term meaning a 'coming
together', or confluence, as when rivers meet. A *sangam*
also means a church, not in an institutional, organized sense,
but in the sense of the early church when Jesus said 'when
two or three are gathered together in My name, then I am
also'.

The *sangams* that have developed are made up of the poorest
village women from the class of agricultural labourers; people
who are so poor that if they do not work, they do not eat.
They are from the caste of Harijans, once known as the
Untouchables. The people that we work with are thus triply
oppressed, by poverty, social status and, particularly, by their
sex.

These women were once invisible and voiceless, but within
the context of the *sangam* organizations they have achieved
many remarkable things. Without access to funds from the
vast aid organizations, such as the World Bank, they have
created a mechanism for coming together to form mutually
supportive asociations. This development has resulted in a
whole chain of advantages for the local people, which are
outlined briefly below.

A community-controlled banking system   The associations started to save their own money. They now control, in total, well over a million rupees, thus creating a non-formal banking system which they themselves manage. Most members are illiterate, but they have sharp memories and they are wise.

The banking system is highly successful with few defaults. If a member does default on a loan the community understands the reasons, and makes allowances. Community banking frees villages from the bureaucratic operations involved in commercial banking and the problems of raising collateral. It also frees them from the temptations of the money-lenders, who usually charge 24–36 per cent interest and up to 150 per cent in a bad year. This leads to the system of bonded labour, semi-slavery, where the family has to work for the rich for several years to pay off a loan.

Creating jobs for the poorest people   The *sangams* have used their money to create employment for their members. When there is no other work available, the women start environmental regeneration projects. They build micro-water harvesting systems and plant trees. They have now planted over two million trees, not Eucalyptus to feed the paper industry but wild fruit trees to help feed their children. These trees replaced those previously cut down as 'not economic' in the interests of 'development'.

These initiatives have also led to increases in wages, and have helped produce new jobs, by creating competitive sources of employment. In these villages, wages are almost double those of surrounding villages.

Maintaining the freedom of the lowest castes   The third major benefit of the *sangam* system is that of political and social freedom, in particular the ability to avoid falling into bondage. In bondage, a person loses a sense of self dignity. If you meet your landlord when you are riding a bicycle, you must get off. If you wear glasses, this is considered to be imitating the upper classes, and you must remove them. You must take off leather sandals. Wearing such simple items sends out social signals that are seen as being a challenge to authority.

This system is based on power. A handful of people control several thousand. There are fewer and fewer human rights as you go down the social scale. The landlord can have any sexual favour he demands as a matter of right. If you poach a tiger

it is a grave offence, because the government is bound
by international agreements to protect the tiger. But if a
policeman shoots ten indigenous tribal people it is not really
an offence, as the policeman will say he did it in self defence,
because the tribal people were carrying sticks. Well, all tribal
people carry sticks.

Therefore, any system that prevents people being caught in
the trap of bondage can play a large part in increasing the
quality of their lives.

Building homes   The women of the *sangams* have also
started building their own homes, using traditional materials,
designs and technology. For the first time in their history the
people are not living in temporary thatched huts. This is being
achieved at a fraction of the cost that the government would
incur if it built the houses.

Creating a health system   The women in the *sangam* have
also created a health system; again, like human rights, a health
system does not usually reach the poor. If a poor person falls
ill, she or he has to travel a long distance to a modern
hospital where they are usually just shouted at, frightened and
confused.

Incidentally, shouting at people lower down the social class
system is a problem in India as in many other countries. One
of the first experiences I had as a student in Britain was being
shouted at in the immigration office. There is a general belief
that if you shout, the person being shouted at will understand
better; this is not the case, they just get rattled. It was only
then that I understood that my shouting at tribals in India did
no good at all. I started understanding my role as oppressor
when I was a student in the West. It is very important that
one sometimes changes roles, otherwise one never really
understands one's own ignorance.

A new form of agriculture   The system of permaculture has
also been established. Bill Mollison has visited the villages,
and three national workshops have been held. There is now a
Permaculture Association of India, with 60 groups throughout
the country.

A non-formal education system   The villages have also
started a non-formal education system for the children.

110

Illiteracy in the rural areas runs at 90 per cent, among the very poorest it is virtually a hundred per cent. People simply cannot afford to send their children to school. They need them to work in the fields, and the half rupee a day that the children earn is important.

So the school system that has evolved is a learning and earning system. The children learn about things like permaculture and organic farming, the relationships between trees, soil and water, land, livestock, people, cycles and systems. They work on the land themselves. They also have modules in Indian society, history, basic maths and science.

One of the children at the school is popularly known as a permaculture expert. He is a boy of thirteen, and he gives lectures to visitors at the permaculture farms. The children come from families that have never gone to a modern school, but now they ask to do so and when we send them they often do better than the children from privileged families.

The role of women

If given a chance, people are willing to help themselves. There is creativity and power amongst the people. The systems described above do not need an enormous amount of funding. The bulk of the money has come from the people themselves. A lot of the effort has come from people giving voluntary labour, for building homes, planting trees, improving water systems and in non-formal education projects.

The women are now thinking of federating themselves into a larger organization. It is significant that the developments discussed above have been carried out by women. I think that the process of colonization and modernization has smashed the psyche of men in India, who have grown up in a patriarchal mode of thinking. They see themselves as people without a future, they cannot look forward to a regular job. They have been brought up to think of themselves as providers, in a world where they cannot be providers. Many end up as drunken layabouts.

The women have a very different long-term view. One woman, when asked why she felt different, said:

When my man earns some money he will first get drunk; after some time he will go and see some woman in the town, and then he will buy himself something he doesn't need, like a synthetic fibre shirt, and then he will look around to buy a bag of grain for the family.

When I earn money, the very first thing that I do is to go and buy grain, the cheapest grain possible, because I need to save money. Tomorrow my husband may be drunk, or ill, or in gaol, or my children may be sick, or the roof may have fallen in, or a neighbour may need some of the money.

One woman's struggle
One of the women leaders was beaten every day for a year by her husband for working with the community. He called her names, threw her about and kicked her. She refused to leave the house, saying that she had done nothing wrong and was simply working for the community. His ego could not stand her independence and he drank pesticide in the middle of the night. His wife ran to the landlord and borrowed his tractor and driver to take her husband to the hospital. She had to lie about how he came to have drunk poison, so that he would not be arrested. The man survived and his wife then told him that she was leaving. He asked her to stay and since the crisis has been a very changed man; he is now even saving money for his daughter's wedding.

Fighting for the forest
Another group of women got hold of a bare hillside to start a forest plantation. The landlords went there with sticks and armed men. The women were defenceless, but sat down and asked the landlords if they were prepared to beat sitting women. The men hung around, threatened and verbally abused the women but eventually went away without doing anything. The women have since grown thousands of trees. The forest shows both the enormous regenerative power of nature and the strength of the community. Anyone who has worked in forestry knows that almost half the cost of managing a plantation is taken up with fencing. Here there is no fence at all, or at least only a social fence, that is the agreement of the community not to allow animals to stray onto the plantation. Along with the trees, herbs and grasses also grow. The grasses are used for thatching huts and feeding cattle. The surplus is simply left on the roadside for anyone to use.

Moving towards sustainable agriculture

The project has also been increasingly involved in working on traditional and other non-intensive agricultural systems.

Community pest control

A particularly voracious farm pest, the caterpillar *amsacta albistriga*, has come to prominence in the last twenty years, due to the indiscriminate use of pesticides, which have destroyed its natural predators, along with other insects, so that it has room to expand its population. It is now the worst rainy season pest, capable of devastating crops. Increasing pesticide levels even further has not helped solve the problem.

With the help of an eminent agricultural scientist we have visited many villages and compiled some of the traditional technology of two or three generations ago to deal with the pest, using cultural methods rather than pesticides. Two methods have been successfully introduced. In one, an alternative food source is provided, by planting a non-crop species, which the caterpillars prefer, on the bounds of the fields or on wastelands. This draws the pest away from the crops. Second, the habits of the moths are studied and pest control methods are used to restrict breeding. The months emerge in peaks, then mate within 24 hours and lay their eggs within 48 hours of mating. In the old days, farmers used to be more in harmony with nature, and knew when to expect the emergence of the moths. When the moths started to appear, the farmers would light bonfires which would attract the months and eliminate 70 per cent of the problem at one time. Groups of voluntary agencies have been co-operating with farmers and the government to start a three-year project using this traditional method of controlling the moths. In some villages, use of this system has resulted in a doubling of income through increased crop production. An ecological approach has also made economic sense. The government was concerned about how to disseminate the idea. But in practice the farmers are beginning to teach each other.

Village seed banks

Along with low input methods of pest control has come the question of farmers growing their own seed. However, this is difficult for open pollinated crops, because of the impact of the wind. Here, group action is important. Crop scientists claimed that it was not possible for untrained farmers to select the best plants, and were amazed to find that farmers were in fact doing so perfectly successfully without a degree in agriculture!

We found out subsequently that the system of seed collection was well known to the old women in the village. Their science was just as good as that of the Indian Council of Agricultural Research, the only difference being in the women's management system, which was much better!

Other examples
Other similar discrepancies between research institutions and working farmers have appeared. Old knowledge has been used to make a number of other improvements, including water saving management of dry-sown rice paddys, rotation of crops, intercropping and water and soil control through simple techniques. Research institutions have only recently realized the importance of intercropping, for example, but the government agriculture department has yet to recognize its importance, despite the fact that farmers have been using intercropping systems successfully for centuries.

We might ask ourselves how these ideas ever came to be rejected as those of stupid old women?

An overview of the development process

In summary: the process described above is one of local action, in which women play a key role:

o *sangams* are created;
o these in turn create local employment;
o members undertake programmes of environmental restoration;
o their efforts help create savings;
o these in turn lead to improved, locally controlled banking systems and improved productivity of the land;
o this results in an increased social status for the poor in villages.

The process also combines use of traditional knowledge, community education, and community participation. The development process is key to the position and rights of women in the community. It must involve genuine people participation, and a process of working together, to form a

114

forest plantation, to care for the soil and so on. Until now, in modern India, as elsewhere, science has taken place in the laboratory and has been tested in field stations. Now, for the first time, scientists and farmers are learning from each other. Scientific research and agricultural extension are no longer seen as separate entities.

Such community action is, of course, a part of Mahatma Gandhi's vision of economic development at the village level, and genuine village independence.

Five problems facing the South

It is worth seeing how the five problems identified at the beginning of this chapter – population, debt migration, environmental degradation and land rights – affect the villages practising this kind of development.

Population   The status of women is a crucial factor here. All the thousands of rural women that I have talked to do not want a large family. They are not thinking nationally or globally so much as that they simply want to be able to give their children good living standards and they know they can only support small families. In areas where the status of women is high, infant mortality rates and birth rates have both fallen. This is not a function of technology so much as a function of the status of women.

Debt   The question of debt has been successfully tackled in these villages; the people are no longer in debt to money-lenders and create and manage their own banking systems.

Migration   People in villages where life is improving do not want to leave their homes. There is a link between the land and their own lives.

Environment   The environment is being regenerated and the productive capacity of the land is improving. This is a long-term process, but there is no finality in nature. The land can always recover, and increase its carrying capacity.

Land rights   The 'trickle down' idea that the rich would give the poor land has not materialized. The revolutionary principle that the rich will be killed and the land taken by the poor has also not happened. But the women of the *sangams* have moved in a lateral direction. There is as much waste land in India as is cultivated, well over 100–150m ha. This land could be cultivated. It is such cheap and unwanted land that the women are taking it on long-term leases; the kind of land that farmers who are interested in agribusiness are not concerned with. By multicropping, green manuring and good husbandry productivity is improving.

Much depends on the status of women, and the status of all the poorest people in the villages. But the improvement of the status of women is a political question, and a question of power and who controls power. Can the rich accept the erosion of their power towards the people, such as farmers, women and children? An erosion of power with honour?

Philippines: Land Reform and Soil Conservation*

ROMY TIONGCO

Manong Magno was a young bulldozer operator when he first went to Damulog, southern Philippines, in search of work in the 1950s. Impressed by the abundant harvests, he decided to become a farmer, and wrote home to his relatives and neighbours, suggesting they join him. So did scores of other settlers in the area.

The available plots were on hillsides, because the best land has been taken over by a handful of wealthy people and transnational companies growing crops for export, such as pineapples, sugar and bananas. The loss of soil down slopes quickly became serious. Even when yields began to fall, from 60–80 bags of maize a hectare to 40 bags, the settlers were too concerned with the problems of the present to worry about the future. Long-term thinking was also inhibited by their lack of secure land tenure. A few NGOs tried to warn them about the dangers of erosion from the slopes, but during the dictatorship of Ferdinand Marcos, NGOs doing community work and advocating social change were labelled subversive and communist. Arrests, cases of torture and disappearances of community organizers and farmers' leaders scared people from listening to them.

After Marcos' overthrow, output in some places fell to 10–15 bags of maize a hectare. The settlers began to drift away. In Maican village, half the farmers abandoned their land. In Kinapat, 80 per cent of the men spent most of the week in another village, returning only at weekends. Farmers had a vivid explanation for their decision to leave land: 'It is possible for us to avoid bullets, but not hunger. When the bones of the land appear, farmers must disappear'. The topsoil gone, they could not grow food on rocks and stones to feed their families.

---

* This chapter has been reprinted with kind permission from *Panascope* (*PS*), the magazine of the Panos Institute, No. 28, January 1992.

By then, however, the Green Revolution programme had so distorted the perception of government extension workers that government reports attributed low productivity to the failure of small farmers to adopt modern technology. Farmers, too, took on this view and scrambled for loans to buy high-yielding seed varieties (HYVs) and ferilizer.

For two years BISAP (Bukidnon Integrated Services Assistance Programme, a development NGO) urged farmers to adopt soil-conservation practices. It convinced only three farmers.

Learning instead of teaching

In 1985 the agency changed strategy. Instead of telling the villagers to stop soil erosion, the fieldworkers asked questions. They got the early settlers to tell the story of their migration to the area. The stories were collated and fed back in visual presentations. The result was dramatic. Almost everyone saw links between settlement, soil erosion and low productivity.

The technique was refined, and tried in another village. This time, however, villagers were trained to ask the questions, collect and analyse the information and present the results to the rest of the community, in a procedure known as the Community Information and Planning System. All 21 farmers who answered the question later underwent training in soil conservation. A similar programme was launched in five other villages. At the end of two years, 272 farmers had contoured their hillside farms with nitrogen-fixing trees.

Gelio was the first farmer to use SALT, the acronym for sloping agriculture land technology. In three years he was harvesting 40 bags of maize instead of ten. The technique is straightforward. The farmers determine the contour lines of a hill with the use of a plumbline and three sticks formed into the shape of the letter A. Two farrows half a metre apart are dug along each contour line. Leguminous tree seeds are planted an inch apart along the farrows. The trees form hedges, preventing topsoil erosion. Every six months the hedges are trimmed one metre from the ground and the cut leaves and twigs are left to become compost. The trees provide the soil with nitrogen, compost is available all year, erosion is halted and eventually the land becomes terraced.

Rodolfo Domingo was persuaded to try the method when he asked a BISAP worker for a loan in order to get his sick wife to a doctor. Four years later he returned to thank the fieldworker for 'twisting his arm', because for the first time since borrowing the money he was able to feed his family from his once-eroded and virtually abandoned farm.

He invited a BISAP evaluation team to the farm. As they looked at the hedges, Domingo turned over a stone and pointed to an earthworm wriggling out of the ground. 'For years I have not found earthworms on my farm', he said. 'Ever since I contoured my farm with these leguminous trees, the worms have come back. My farm is becoming fertile again.'

119

Thailand: Land Security and Traditional Agriculture

LARRY LOHMANN

It is important to start by stressing the limitations of a presentation of this type and length. Of course, no one person can every speak for a country, and this becomes even more true when we are considering the complexity of the problems connected with sustainable agriculture and land rights in a region of the size and diversity of south-east Asia.

In the Philippines, for example, there is a long-standing and very polarized land conflict relating to the old plantation economy. This is related to sustainable agriculture problems in that landless people are forced into marginal areas in the hills, a situation which is repeated in many other countries. In Indonesia, problems are again partly related to the colonial legacy, and more recently have resulted from transmigration, where, in accordance with a government-sponsored project supported by many international aid agencies, people are moved from the inner to the outer islands. The result has been human degradation, misery and, from an environmental perspective, loss of soil. To take a third, very different example, in Papua New Guinea over 90 per cent of the land is owned by people's landowning associations, but problems of exploitation, trickery and degradation still occur, and landowning associations are constantly drawn into conflict with loggers and others.

This chapter will concentrate on Thailand, and particularly on the subject of local people's control over land, and of their resistance to the expropriation of this land and of other resources. These matters are closely linked. In Thailand, the issue of defending land is related closely to defending ecological or organic agriculture.

The history of land control in Thailand

Historically, the economization of land (that is, drawing land into the market economy) is tied closely both to the destruction of traditional agriculture and to the development

120

of new forms of agriculture. On the positive side, experience suggests that the resistance to the takeover of land by the state and commercial forces is often strongest in areas with long-standing and sustainable forms of agriculture. Conversely, the proliferation of some newer or rediscovered practices of organic agriculture in Thailand becomes easier when local people's access to land and other connected resources such as water and trees is secure.

In the mid-nineteenth century, the Thai government made deals with colonial powers, particularly Great Britain, ensuring that there would be no political takeover of Thailand so long as the élites promoted the expansion of rice production for export to the colonies. This allowed the élite to expand its control over the country, and the colonial powers to obtain cheap rice for their colonies.

Around the 1890s, such deals were extended to other commodities. British timber companies were given concessions by the government to carry out logging in the north, taking land away from local people. Again, this enabled the élite to expand its own control, this time over areas in the north of the country.

The result of these deals was a change in the economic status of the land. In rice-growing areas, land began to acquire an economic value which it did not have before. It changed from land where people had a common right to cultivate areas that they cleared themselves to land with property rights, and a price. In the north, areas which had been used by villagers as forest commons for hunting, collection of wood, vegetables, mushrooms, bamboo etc. were turned over to the state and to corporations, resulting in loss of timber resources, viable water sources, and so on.

This latter process, which continues today, has led to a shift in attitudes towards the land. As common property regimes are replaced by 'open access' regimes, local people reason that they might as well grab whatever resources they can from an area before it is taken away from them by outside forces.

Further changes in the twentieth century

The takeover of common land and forest rights accelerated in the twentieth century, with the beginning of the era of so-called development after World War Two. World Bank and United States government missions came into the country and began to encourage the élites to integrate themselves more fully into the world economy. This resulted in a rapid escalation of change. To transport crops out of farming areas for export, roads had to be planned, financed and built. The same roads were used to distribute consumer goods and agricultural inputs around the country. Industrialization of the cities required electrical power, which meant that rivers were taken over and dammed to develop a hydroelectric generating capacity.

The local élites benefited from these changes in many ways. Roads into the interior allowed them to increase land speculation, or obtain government logging concessions. Equally, it allowed the central government to consolidate and increase its control over hitherto remote regions. Industrialized countries benefitted from an increase in export crops, promoted by the World Bank and the Thai élite, such as cassava, corn, sugar cane and coffee. These crops are not traditional crops. They are upland crops, and encourage the clearing of upland areas. Such clearance has resulted in serious damage, including erosion of soil and degradation of water stocks. This affects the ecology of both the uplands and the lowlands.

The new agriculture undermined traditional ways of life. New inputs were required, high yielding varieties (HYVs) of seeds, pesticides and fertilizers, in order to grow the new crops. Concurrently, the traditional need for status now had to be met by consumer goods. The overall result was debt, loss of land, migration, forest clearance, and the increasing destruction of ecosystems and degradation of neighbouring land.

It is important to note here that control of land is not just a legal matter of title, but of retaining other necessary resources as well. If water is taken away or degraded by someone growing cassava further up a hillside, then it becomes even more difficult to hold onto land traditionally controlled by a family or village. If new roads and business come into a village, the people controlling these will tend to make alliances with village leaders, reducing the latter's accountability to the rest of the villagers, and thus further reducing the powers of the poorest people.

Villagers are, on the whole, well aware of the loss of old systems, and the negative aspects of modern developments. Their problem is one of control and the difficulty they have in fighting or adapting to these changes.

The situation today

The extension of control of the market and state control into marginal areas has now reached a new stage, and a new wave of expropriations is taking place. The government, which previously saw marginal areas as being able to absorb some of the people who had been driven off more profitable land, is now encouraging market forces to take more direct control of these areas. New commercial opportunities are emerging, notably for Eucalyptus plantations, which are an important source of raw material to the paper pulp industry. In some areas, salt mining is also important, but has serious side-effects by polluting water and thus impeding rice cultivation.

The taking of villager's land is encouraged by the granting of government concessions to private firms and informal sales of land to local notables. The government forestry department had been given control over about half the country's land area by the 1960s, leaving all those living on the land without prospect of title, despite the fact that they felt as if they should have entitlement under common law. This has led to an explosive social and political situation, with frequent confrontations between villagers and commercial and state interests. This atmosphere is not conducive to such long-term aims as the development or recovery or organic agriculture.

An experience of an NGO working with villagers in the north helps illustrate the problems facing anyone trying to develop sustainable farming systems under these conditions. The area in question was a part of a National Reserve Forest, but was occupied *de facto* by villagers. The NGO worked with the local villagers, organizing exchanges and providing advice in an attempt to come up with new approaches to dealing with the ecological and economic difficulties posed by farming on steep hill slopes. For example, some villagers were trying to substitute for nitrogen ferilizers by planting nitrogen-fixing trees in contours along hillsides, which over time helped build up natural terraces, reducing erosion and the need to clear more land. The project was proceeding well, but then a local doctor decided that he wanted to buy up the area to develop a resort for rich people from Bangkok and Chiang Mai. Although frightened of losing their land, the villagers were also attracted to the ready money offered, and the income that the resort might generate. The sustainable farming project is continuing due only to the ability of the NGO to forestall the land grab by buying some of the land itself.

123

On the positive side, the connection between the role of sustainable agriculture and the control of land is also shown by the fact that resistance to expropriation of land is often strongest in areas where traditional organic agriculture is practised as part of the culture.

This connection is clearly demonstrated by the traditional irrigation systems in the north of the country, which have been used for many hundred of years. The hill forest provides the sources of streams and also provides other goods for the villagers. Water is carefully managed, so that it feeds the rice fields both on the hillside and in the valleys. This system does not distinguish between agriculture, forestry and water management; all three are essential to each other.

However, when the modern agricultural system was introduced, the forests were given to logging companies to help contribute towards national income; streams were given to the power companies; and agriculture was pushed toward producing cash crops. Obviously this was a threat to the villagers' livelihoods. Taking away the forest meant that the streams were also removed from common use and silted and dried up in the dry season. Damming the streams for power meant that the water could no longer be controlled for the agricultural system. The villagers saw clearly that a threat to any part of the system was a direct threat to themselves, their way of life, diet, culture and so on.

Resistance was strong. The logging ban introduced in 1989 was in part the result of pressure from villagers in areas like these whose traditional irrigation practices were threatened. Villagers took part in marches, blocked logging roads, prevented companies from taking logs out of the forest, demonstrated at local district centres, compiled petitions, demanded to see the prime minister and joined forces with villagers from their own and other regions of the country. The government was forced reluctantly to begin to acknowledge the legitimacy of village claims.

Again, this demonstrates close connections between control of land and the forms of agriculture practised. The development of organic agriculture in countries such as Thailand can only occur when the land and the culture which regards agriculture as a way of life are both secured.

The practice of organic agriculture

Farmers in many regions have worked out appropriate systems of organic agriculture for themselves, for example, systems of mixed farming, including intercropping, fish ponds, fruit trees, rice paddies and pigs. Some farmers have spent thirty years of their lives working out how to do this in a way which respects the local environment and maintains their independence and culture. This type of agriculture is only found in areas where villagers feel secure in their land tenure. These villagers have not rejected the market totally, but they have consciously put it into a subordinate place and often wish to keep government officials at arms' length.

How is land security to be achieved?

Given the problems and pressures outlined above, how is land security to be achieved in practice? How are local people to maintain or regain power over local land, water and forests, to say nothing of credit, capital and political connections? Is there anything that large foreign or international agencies can do to help? Two quite different examples may serve as a background to further discussion.

The Tropical Forestry Action Plan

Thai NGOs already have some experience with the Tropical Forestry Action Plan, the environmental Marshall Plan of the present day, being promoted by the Food and Agricultural Organisation and other international bodies. The Finnish bilateral aid agency (FINNIDA) has been entrusted with the financing of this plan, and has hired a Finnish forest consultancy company to draw up a strategy to save the country's remaining forests.

Thai NGOs saw a preliminary draft of the proposal and discovered that the recommendations were 80 per cent for industrial forestry, for example, planting Eucalyptus for the paper industry. The NGOs tried to exercise political influence to change the plan; they held a public meeting of all the parties in the hope of creating some changes. The Finnish consultancy was taken aback by the opposition from NGOs and villagers and promised to revise the plan. However, the changes have not been forthcoming and some Thai NGOs have asked for the scheme to be abandoned. The problem in Thailand, as in other countries, is that large aid bodies are still working with the élites in their host countries in opposition to the people affected most directly by the plans who have no say in their formulation.

The opposition to the Pak Mun dam in the north east
Another example of the intransigence of aid agencies is shown by developments with regards to hydropower. In northeastern Thailand, villagers have practised agriculture along river banks for centuries, using silt from the Mun river as a fertilizer. This system is now threatened over a wide area by the Pak Mun dam, now under construction. In 1991 twelve thousand local people petitioned against the dam, and presented their fears to the World Bank, which was proposing to part-finance the development. The villagers have confronted the Bank directly both with their scepticism about the supposed benefits of the project, and with questions about the Bank's intervention in projects of this type.

They pointed out that if the dam were built, they would be forced to move away from their traditional homes to infertile lands. These lands have already been abandoned by most of their previous tenants and are unacceptable to the villagers.

Many Western NGOs subsequently joined the villagers in opposition to the project, and in the attempts to convince the World Bank not to sponsor the development. However, the World Bank has rejected the claims of negative aspects of the project, and sees the issue as one of finding the right price to pay off the villagers. On 10 December 1991 the Bank's Board of Directors voted to approve the loan.

In both the examples given above, the villagers have assessed the effects of major institutions on the local way of life and have decided that this path of development is not for them. Instead, they want to retain control over their land and their lives.

To finish this presentation, it is pertinent to relate a response made by a Kenyan delegate at a recent NGO meeting on UNCED in Bonn, which sums up the attitudes towards foreign intervention in 'development' held by many people in the South:

I have heard a lot about how many sacrifices the South is going to have to make for the sake of the world environment; but I would like to ask a question about the sacrifices the North can

126

make for Africa. Do you dare in the North to leave Africa alone for ten or twenty years to work out its own solutions? Do you dare to *stop* the financial flows not only from south to North but from North to South? *Do you dare to stop your so-called aid?*

Indonesia: Problems of Land-use Policy

TADJUDIN DJUHENDI

The present state of both land-use and agricultural practices in Indonesia can be traced back directly to the
country's framework of economic development. First, sectoral
development has been significantly influenced by the ideas
of Rostow and Rosenstein-Rodah. Satisfying national needs
by producing more food, and various 'modern' products,
underlay the national spirit of development in the late 1980s.
The development trend was then shifted into industrialization, in order to substitute for import commodities.
Finally, the industrial and agricultural sectors were both
strengthened so that more export commodities could be
produced.

It is also necessary to note that regional development
has been exphasized in Western Indonesia, especially in
Java, since this island is more suitable for industrial development (see Table 1). This has important implications in
that Java has undergone rapid growth in both its industrial
sector and the size of its cities, thus explaining how it
has come to be Indonesia's centre of commerce, state affairs
and industry.

Table 1   Areas and population of Java and Outer Java Islands, 1989

Description		Java	Outer Java
1.	Area (km2)	132,187	1,787,256
		(6.89)	(93.11)
2.	Population	107,513,798	71,622,312
		(59,81)	(40.19)
3.	Population density per km2	826	140

It is important to emphasize this general background,
since it has far-reaching consequences for land-use issues.
Furthermore, it is virtually impossible to discuss sustainable
agriculture without also considering general land-use problems.

128

Conversion of agricultural land

In the economic climate described above, the industrial sector, along with other urban sectors, gained a leading position. This has given these sectors a powerful bargaining position with which to obtain more space to continue their growth.

Increasing industrialization and urban growth has immediate effects on agricultural land, which has been reduced in area. This process of conversion of land away from agriculture has been most intensive on Java (see Figure 1).

The impact of this conversion has not always been quantitative, in terms of farmed land lost, but it has implications also for the quality of agricultural land. To compensate for the

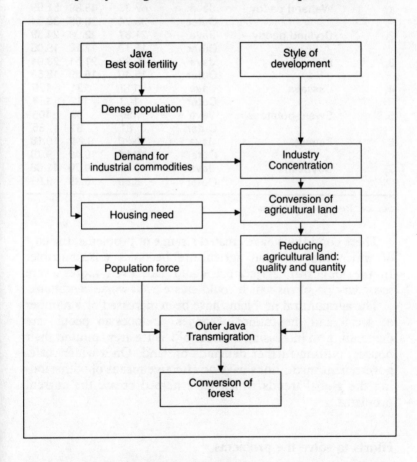

Figure 1 *General mechanism of conversion of agriculture land (Based on Java–Outer Java Island Relationship)*

129

decreasing area of farmland in Java, the government has attempted to create new agricultural areas in the outer islands, through the transmigration programme and paddy fields (*sawah*) construction. However, the quality of the new farmland is generally lower than that of Java, both in terms of its natural fertility and its soil structure. These differences, and the resulting degradation of quality, are shown in Table 2 and Figure 2.

Table 2   Food Crops Productivity in Indonesia (000kg/ha)

Crop		Location	1985	1987	1989
1.	Wetland paddy	Java	47.27	48.89	51.55
		Outer	35.74	36.05	38.34
2.	Dryland paddy	Java	21.87	22.98	24.38
		Outer	15.73	17.10	19.05
3.	Maize	Java	19.23	21.51	23.04
		Outer	15.40	16.89	18.50
4.	Cassava	Java	112	121	126
		Outer	103	112	116
5.	Sweet potatoes	Java	90	98	103
		Outer	81	81	85
6.	Peanuts	Java	10.20	9.24	10.18
		Outer	10.64	10.40	9.78
7.	Soybeans	Java	10.19	10.70	11.68
		Outer	8.80	10.35	9.99

Source: Bureau of Statistics

These conditions have created a range of problems, the cost of which may be that agriculture becomes unsustainable. In the long run, agriculture could face even more severe secondary problems which could create even worse conditions.

The agricultural problems have been increased by a number of social and institutional changes. Indonesian people are demanding an increasing number of leisure areas around their houses, putting further demands on land. On a wider scale, government institutions have no effective means of withstanding the global trends which have helped create the current problems.

Efforts to solve the problems

Sustainable agriculture does not only rely on efficient agricultural practices, but is also influenced by land-use policies.

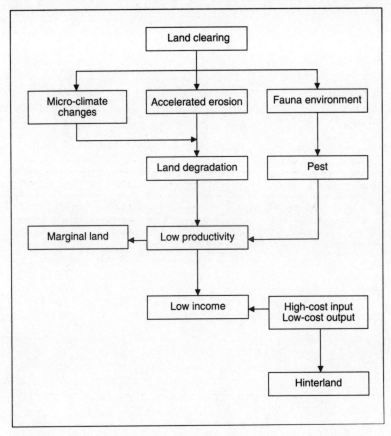

Figure 2 *Pattern of land degradation due to the resettlement programme in Outer Java*

This means that Indonesia, along with other countries in the South, needs effective legislation to prevent damaging imports from improper land utilization, in addition to the maintenance of agricultural practices, such as soil conservation, minimum tillage etc. Such legislation should concentrate on two key issues: suitability of land for agriculture, and controlling the growth of urban areas.

PART 5: *Europe*

Organic Farming in Western Europe — Which Way Forward?

LAWRENCE WOODWARD

Since the 1970s the primary goal of the majority of organic organizations in Europe, and the USA, has been to promote the terms organic, biological or ecological and the practice of organic farming in general. The aim has been to increase the number of organic farmers and organic consumers and to establish the term organic on the agendas of agricultural researchers and advisors and with those involved in agricultural marketing, training and policy.

Throughout Europe the number of organic farmers, number of hectares under an organic system, and funds for organic research remain low in comparison with conventional agriculture. It is clear, however, that matters organic are now, although not 'centre stage', at least 'bit players'.

By 1990 there were 10,000 organic farms within the 12 countries of the European Community (EC), farming 161,000 hectares of organic land. In non-EC countries there are a further 4400 organic farms (1250 in Austria, 850 in Finland, 300 in Norway and 2000 in Sweden).

In market terms, in countries such as Germany, the United Kingdom and Denmark between 1–4 per cent of the retail food market is in organic produce. In Denmark 8 per cent of milk sold is organic. In financial terms the value of the organic food market in the UK in 1991 was approximately £120 million and the trade deficit in organic foods is a cause for concern for the government.

Throughout Europe farmers are beginning to receive support for converting to an organic system. In Denmark, Sweden, Norway, Finland, Switzerland, Austria, Spain, the Netherlands and several of the German states financial support is available for conversion. Developments are also taking place in Eastern Europe. Poland and Czechoslovakia, for instance, both now have ministers for alternative agriculture.

An EC regulation which will have become operative in June 1993 provides a legal definition for the term 'organic' and labelling of organic foodstuffs. Several EC research programmes have indicated a willingness to support organic research, and an appendix in the EC agriculture minister's proposals for the reform of the Common Agricultural Policy (CAP) include references to organic agriculture.

The problem of maintaining standards

For those of us who have been campaigning for the uptake of organic agriculture for the last decade these figures show some measure of success. But at what price?

Is the form of organic agriculture which is now being taken up and enshrined in legislation quite the sustainable production system we envisaged? Or have the definitions and practices of sustainable agriculture been changed in the interim?

There are developing several forms of agricultural systems held up to be 'sustainable' and 'environmentally friendly' which seem to have little in common with the vision of sustainable production systems originally campaigned for by these activists. In the USA, for example, a system of production called LISA, Low-input Sustainable Agriculture, is being promoted. Surely, low-input systems are tinkering with conventional systems and having nothing to do with sustainable systems? Similarly in the UK, a recent initiative by the Ministry of Agriculture on sustainable agriculture was not seen as involving organic agriculture but rather as modifying conventional systems.

Many of the marketing initiatives involving organic food, especially in the UK, have been developed by large multinational companies using conventional models of production, distribution and retailing. They have recognized that if you use the term 'organic' you can add value to a mature or static food market. There are, therefore, many large farms in the UK producing, processing and distributing food using the conventional model but in accordance with organic standards.

Britain also has a large market for imported organic foods. There are all kinds of so-called organic foodstuffs produced within conventional production, distribution and retailing models being exported to Germany and the UK from the USA. The UK also imports tea from a large organically run tea plantation which is owned and run on conventional models by the large multinational Lonrho, and 'Cheddar' cheese from a large processing scheme in Poland.

Once in force the EC regulation will govern, and set the standards and context, for organic trade in Europe. It emphasizes only organic techniques and says nothing about social or ecological perspectives; it has no wider vision.

Equally, far too many of the government-funded research projects in western Europe and the USA are based on a view of organic production as a niche market not as a mainstream option tackling problems of environment, ecology, food quality, animal welfare or land reform.

A warning

It may be that these views are over pessimistic, and clearly there are exceptions, but we should recognize the real danger, especially in the UK, of organic agriculture being taken over and remodelled by the agro-industrial complexes. The value of organic agriculture to movements such as that for the reform of land-use is in danger of being lost.

Organic agriculture techniques are clearly and demonstratively the best methods to achieve really sustainable agricultural production. They are the best methods to achieve stability in rural areas, to protect the environment and regenerate derelict land. These benefits will, however, be lost if we concentrate only on techniques.

IFOAM

These wider principles are clear within the objectives of IFOAM (the International Federation of Organic Agriculture Movements), which aims to unite the efforts of its members to promote organic agriculture as an ecologically, socially sound and sustainable method of food production. They are principles that aim to achieve self-sufficiency, decentralization of production and distribution, and the encouragement of human values and potential. Without these principles organic techniques are in danger of being worthless.

In 1989 at the conclusion of the 7th International IFOAM Conference in Ouagadougou, Burkina Faso,

135

participants drew up the *Ouagadougou Appeal* which states:

Considering:

○ the serious food deficit in many developing countries;
○ the disturbing nutritional situation for large numbers of the inhabitants;
○ the rapidity and extent of the erosion of soils, desertification and pollution of the environment the organizers and the 600 participants from 48 countries declare:
1. That for developing countries ecological agriculture is not an alternative but a necessity imposed by local conditions.
2. That there is a need to take into account agro-ecological practices in attaining food self-sufficiency.

What can we do?

If we wish to ensure that the organic movement and sustainable agriculture remain on the same track we will have to ensure that the organic organizations take on these wider principles and fight for them within the regulations and laws to define organic production. We will have to stop consumers buying organic produce just becaue it is labelled organic and educate them to buy food produced in line with these wider objectives.

Those who wish to encourage, support and bring about land reform, as well as those within the organic movement, will have to fight for and win back the term and concept of sustainability. We can no longer put up with the degeneration of the term. And we must continue and renew our opposition against the GATT (General Agreement on Tariff and Trade) agreement to establish free trade, exploitation and unsustainability.

The Norwegian GATT Campaign

HELGE CHRISTE

Norway has only 50 per cent self-sufficiency in food and it imports 60 per cent of staple wheat. This means that its food requirements are very different to those of most other EC countries, and more reminiscent of some Third World countries.

There are many Norwegian farmers who would like to see a more sustainable agricultural system, i.e. environmental agriculture, low-input agriculture and traditional agriculture, although there is a variety of opinions about such a development.

Farming groups and environmental organizations have been working together in Norway to prepare responses to recent initiatives such as the latest GATT talks, FAO policy developments and UNCED.

The alternative programme that those groups have drawn up for Norway has nine main points:

○ Each nation should determine its own system of food security and its own food policy.
○ Each nation should have the right to set higher standards on environmental, health and security questions. The GATT proposals include the harmonization of global environmental standards. This does not allow nations to set an environmental standard higher than the agreed level and it does not encourage local and national initiatives to increase environmental standards.
○ The hidden environmental costs should be included in food prices.
○ Use of land is a great responsibility, and land should be controlled by those who cultivate it. In Norway we have a very good system, which obliges land managers to both cultivate land and live on the farm.
○ Awareness of the risks involved in biotechnology. We believe that the test of social needs, the so-called 'fourth hurdle', is an important criteria for biotechnology development.

137

- Pesticide use should be reduced and maybe elimimated. This should also include a global, legally binding code of conduct to prevent illegal export of pesticides, thus stopping the 'circle of poison'.
- Organic agricultural systems should be introduced. We hope all farming communities will make the transition towards sustainable agriculture.
- Third World countries should produce mainly for local needs. Export should only be a secondary aim.
- Debt elimination should be introduced and aid programmes directed towards local needs. We believe that many aid programmes from industrial countries are designed to support industrial countries, rather than the countries receiving aid.

Organic farming in Eastern Europe*

NIGEL DUDLEY

With a note on Estonia by VIIVI ROSENBERG

In 1978, Boris Komarov, a high-ranking Soviet official smuggled out the text of a book published in Britain as *The Destruction of Nature in the Soviet Union*. This provided for many people the first inkling of the scale of environmental problems behind the Iron Curtain, including widespread pollution problems on land:

> The local inhabitants of Uzbekistan never pick the largest water melons or musk melons in the market. Giant melons can now be grown simply by applying huge quantities of fertilizers, particularly saltpetre. Fortunately, saltpetre, which concentrates in the melon pulp, makes them bitter, so that cases of serious poisoning are comparatively rare.

Since then, stories of environmental pollution in eastern Europe have become legion, and concern about green issues was a key factor in many of the democratic revolutions during 1989. However, the green movements in the central European countries remain small, underfunded and desperately in need of information and support. In particular, the green movement organizations and activities often have few responses to the hard sell from the agrochemical companies who are fighting hard to make sure that agriculture in the east follows the same path as in the west. Agriculture has been either regimented in intensive communal farms, or effectively left in the peasant state, as in parts of Poland and the former Yugoslavia.

There is, however, a growing number of organizations involved in producing, certifying and marketing organic food. Much of this comes to the west. These organizations are usually small, but enthusiastic, and they need help from similar groups in the west.

---

* This chapter is an updated version of an article from *Living Earth*, the journal of the Soil Association, Oct–Dec 1990.

139

Czechoslovakia

The Czechoslovakian organic movement is still tiny; most of the members were involved in the political opposition until little more than two years ago and are only now starting to develop practical projects. There is an organization, Bioagra, which ran its first major conference in December 1990. There are currently four organic farms, operating under biodynamic principles. The movement is enthusiastic and determined to expand. The environmental situation in some areas of the country are already critical, with some 5 million tonnes of topsoil being lost every year as a result of erosion.

Hungary

Hungarian agriculture changed radically under the communists. In 1935 there were 1.2 million farms, run mainly by peasants living in appalling privation. By 1979, collectivization had reduced this to just 141 vast state farms, and 1425 co-operatively run farms. In 1986, economic reforms allowed members to hire land and machinery from the co-op, and two years later 11 per cent of the land was already back under private management, growing 22 per cent of the produce.

Currently, there are about 3000ha under organic production, concentrating mainly on wheat (almost a third of the total), sunflowers, soybeans and rye, oats and barley, although with a range of other products. There are 35 farms involved, along with four orchards and four vineyards. Organic production has already had a rather erratic history, with one company going out of business after three years. The BioKultura Klub, which was established in 1986, runs a certification system. However, most of the food goes for export and at present there is apparently no real market for organic foods within the country, both because of lack of awareness about food hazards from intensive agriculture and a shortage of money.

Poland

The agricultural-political situation in Poland is perhaps even more confused than that in Hungary. The Smallholders Party is currently the second largest political party in the country, and is demanding that all state-owned land be handed back to the original owners. This seems unlikely to take place in practice; many are now dead, and their ancestors have moved away from the land. At the moment it is acting mainly as a spoiler for other developments.

There are a few organic producers already operating in Poland, including at least one experimental farm. There is some interest in developing organic farming among communities for the mentally ill, in ways which could be similar to those of the Steiner communities in western European countries.

Romania

Much of Romanian agriculture is effectively organic, in that it is still so undeveloped that there is little use of agrochemicals. However, modern organic techniques are not used, and the agriculture is very inefficient. Government policy continues to be confused and, before the recent troubles, there were drastic moves to intensify agriculture. Due to the continuing political confusion in the country, it is impossible to predict what will happen in the future.

The next few years will be a critical time for agriculture in eastern Europe. Poor organizations, lack of incentives and the failure to adopt some of the more acceptable of the new techniques mean that production falls behind that of the west in many cases. On the other hand, some areas have far fewer of the problems of reliance on agrochemicals than is the case in most other industrialized countries. There is already enormous pressure to intensify, both from within the countries, where people are desperate for exports and foreign currency, and from the chemical companies and governments in the west, which see a potentially huge new market for their produce.

Other issues are involved as well. Imports from eastern Europe could add to the destabilization of our own agriculture. The cash crisis which is crippling the economies of the newly democratized states should add further incentives for developing the least costly methods of agriculture but, as in the Third World, we are likely to see aid money being used largely to bolster our own exports.

Soil and agricultural problems of Estonia

A note by VIIVI ROSENBERG

Estonia, where people have been living for about 5000 years, is considered to have average conditions compared to the world as a whole. Our soil and climatic conditions are worse that those of Holland, Denmark, Central Russia or the Ukraine, but better than those of Spain, Armenia and some places in Africa.

The Estonians have been mostly tillers. Our ancestors lived primarily on agriculture. As often happens to small peoples, the Estonians have not enjoyed much independence. Most of the time we have had to live under the rule of foreigners, and this has influenced our culture and way of life.

At the beginning of the present century the system of small farms was formed. Enough agricultural products were produced for our own people and for export. The success was also due to the fact that each farmer knew the soil of his fields. This was possible only because of their long and settled farming tradition. Farming was handed over to sons who from the earliest childhood learned this job from their fathers.

Estonian agriculture made progress after 1919 when we secured independence, from Russia.

The fact that Estonia exported potatoes and seed potatoes to many European countries such as Holland, Denmark, England, France, Spain, Germany and even to Argentina is evidence of the high level of the Estonian agriculture.

From 1940 to 1991 Estonia was unlawfully occupied and a new experimental system was introduced. The old agricultural system was rearranged. Farm owners became employed workers on big farms which all belonged to the state.

The more successful families lost everything and were deported to Siberia. The big machines were brought in which ploughed our soil in a wrong way, too deep, and made the soil thick. The fertile soil-bed is thin in the Estonian soils, in some places only 15–20cm. Bigger fields (50–150ha) were developed instead of the old (2–5ha) ones. In many places the traditional farm-houses were abolished.

It would take much time to point out all the negative sides of socialist large-scale agriculture. Those people who have not met these problems directly, can hardly understand all this. In short, this agricultural system was far off the economic stand. For example, you could often see the roadsides covered with fertilizers in spring and with grain in autumn. It was a wasteful, witless system which caused a lot of damage to people, animals and the surroundings.

Now we face a complicated, difficult and time-consuming task, namely to restore the good standard of agriculture, in the first place to re-establish the fertility of soil. There is a saying among Estonians: 'If you tease the field once, the field teases you back nine times'.

In spite of the fact that there are lots of problems ahead which need resolving, our scientists don't think only about our future and people of Estonia. Our interests are much wider. We are working for the future of all people of the world.

The aim of the research programme of which I am head (EVIKA) is to develop plant-clones of potatoes which have greater yield potential and are more resistant to diseases. We are trying to find propagation methods which are cheap, simple and don't pollute the environment. We have for instance worked out a method for propagating potato which speeds up the process. Using this method we can get 2–5mm plants from one plant and 40–50mm tubers during a year. This method is cheaper and simpler, than other methods, it needs less energy and does not harm the environment.

By using this technology we could also develop potato breeding in those countries where it was not possible before because of the short growing season. The technology is currently being used in many districts of the Commonwealth of Independent States, in Siberia, Kazakhstan, Russia and also in Latvia and Lithuania.

We have now expanded our research work into fruit and other kinds of trees with the aim of developing technologies that allow for quick and cheap propagation. This should help to establish new gardens and forests and restore the damaged ones.

Viivi Rosenberg

APPENDIX

Statement from participants of the International Conference 'Soil for Life', held in Berlin, 23–7 November 1991

We are deeply concerned that the present way of producing and distributing food is not working. While there are surpluses of food, a billion people go without enough to eat. The present world agricultural system is increasingly unsustainable. It is degrading the resource base and poisoning the environment. We have only considered the narrow economic price and not the true costs in ecological and social terms. Moreover, the promises of the present system have not materialized for the majority of the world's people. Outdated economic theories and misconceived priorities such as urban bias lead inevitably to a situation where a transition to sustainability becomes more difficult the longer we wait.

All the world's people can be fed if the North and the South adopt sustainable agricultural policies and move toward an agricultural system that works with, and does not dominate or deplete, natural systems and resources with external inputs of artificial fertilizers and pesticides and high energy consumption, but builds on local ecology, knowledge and skills. The presentations at this Conference have demonstrated that integrated land-use, through organic agriculture, can produce the food the world needs. It results in higher productivity on small farms and on poor or degraded lands and is competitive with chemical systems once the true costs are taken into account.

The Conference concluded that organic agriculture is best achieved when people have equitable rights to land and other agricultural resources. Thus in many societies this will involve a significant shift in power. We consider humanly unacceptable the monopolizing of the land by a minority of owners, as is the case in Latin America, south and south-east Asia. The Conference supports popular initiatives seeking democratization of land ownership. Agrarian reform and organic agriculture are indispensable conditions for democratic development, for people as well as nature, for the countries of the South as well as those in the North.

The Conference emphasized the importance of social and environmental obligations of land ownership and management.

The Conference recognized that the desirable transformation of agriculture would need appropriate support especially during the transition period. Financial, research, rural training, education in the North as well as the South, among other policies, need to be redesigned.

We are deeply concerned that the direction of biotechnology will have further adverse effects on natural systems and increase the exploitation of the South.

We understand transitions to sustainability will take many forms. Therefore, the Conference rather than prescribing detailed solutions, emphasized the need for strengthening the role of the rural majority and their organizations to make them equal players in this process. Northern aid organizations need to provide adequate resources for this task. This implies, for example, protecting their rights to land they improve and creating new forms of landownership; building awareness especially in the North of the need for social constraints on production and consumption.

Sustainable agriculture and agrarian reform are unlikely to be achieved unless certain international constraints are removed: for example, the foreign debt burden that is impoverishing people and their environment; the IMF structural adjustment programmes which oblige countries to increase exports invariably at additional costs to their environment. In this connection, the North needs to recognize and pay its environmental debt to the South as well as to compensate the South for its role in protecting the global environment. It is, for example, the North that emits most of the CO_2 that damages the atmosphere for all the world's people. In this respect the Conference supports the call of the German Federal Minister for the Environment of a policy of 'debt for debt', whereby the debts of the South are offset by environmental debts of the North.

With respect to UNCED and the Uruguay Round of GATT, the Conference calls on governments to give equal priority to environmental and development issues.

Food security requires a radical revision of international trade and agriculture policies away from export dumping and towards trade relationships based on the principles of fair exchange, including ecological and social criteria.

We reject Northern policies and practices which block

agrarian reforms. We call on people in the North to seek ways of supporting groups working toward agrarian and agricultural reforms in the South.

Group reports

During the Berlin conference, three specialist groups met to consider land reform, sustainable agriculture and food security. This is a summary of their discussions and conclusions.

Land reform

'Land reform' is an inadequate phrase, believed a number of Latin American participants in a specialist group on the issue. 'Agrarian reform' is preferable because it indicates that land reform needs to be accompanied by other reforms. Agrarian reform has to provide access to land, enable people to make use of that land, security of tenure, and the prospect of sustainability. It involves tenure, farming systems and agricultural services, such as credit and extension. It is little use giving people land, unless they can obtain the services they need to work that land.

People in rural areas know the local situation and can make an important contribution to the debate over the best way to achieve agrarian reform, but they are often not taken seriously by government. Agrarian reform is part of the process of democratization, but it cannot take place unless there are the political requisites for it. In some countries, Brazil is an example, power structures are seeking to prevent agrarian reform. Such reform is about power; it will change power structures.

Unless there is agrarian reform, the problem of deforestation will continue, for it is landless people who are driven by desperation into the forest to try to clear land to make a living. Non-implementation of agrarian reform means continuing soil erosion; people who do not own the land they farm may not take care of its soil; they are more likely to pump it for all its worth in the short term.

International solidarity is needed to help the process of agrarian reform. There is a need for solidarity between Northern governments and Third World peoples seeking agrarian reform, and also from growers and concerned groups in the North; the latter could help to raise the level of awareness about the issues. Growing political awareness

147

can lead to change. (Eastern Europe and South Africa, for example.) Farmers in the North may have had experience of agrarian reform which gives them some empathy with people in the South, who are now seeking it.

Governments of countries in the North could help the process of agrarian reform by looking very carefully at their aid policies and ensuring that those policies are not preventing reform. There is a case for some kind of impact assessment of aid. Representatives of land reform organizations should be represented on their country's aid delegations.

The foreign debt problem that now burdens many countries can hinder the agrarian reform process. Indebted countries often lack money to tackle poverty and landlessness. Countries struggling to earn foreign exchange may keep plantations that grow export crops rather than redistribute that land to local people.

There is a case for a world conference on agrarian reform, believed participants, building on the 1979 FAO World Conference on Agrarian Reform and Rural Development, making use of the work that has subsequently been done.

Sustainable agriculture

Agriculture is about more than just producing food, believed a specialist group; it is a system that underpins the whole of our society, our relationships, including our relationship with the environment. There has been an alienation of people from nature, particularly but not exclusively in the North. Combating this alienation from the land is important; what is needed is education in the broadest sense, through schools, the media, and through schemes such as rural development programmes, on-farm training, demonstrations.

If we want to see environmentally balanced sustainable farming systems, we have to tackle political issues, such as the discussions of the Uruguay Round of the General Agreement on Tariffs and Trade concerning agricultural trade between countries of North and South.

Organic agriculture is preferable to industrial agriculture, and it is important not least to see it as something that can reduce the dependency of one country on another and one region on another. It is a matter of debate whether organic agriculture is the sole way forward or whether it is part of a series of different options including low-input agriculture. However it was agreed that farming systems, including organic

148

farming systems, that are introduced into the South should be flexible and adaptable to local conditions, and not imposed from outside.

Among the advantages of organic farming was that it was an approach that maximized re-cycling, minimized inputs from outside, and, if outside inputs were essential, was a system that concentrated on the least damaging methods, both ecologically and to health.

The definition of organic farming needs to be expanded, particularly with respect to farming in the Third World where there are different traditions, and also with respect to wider social issues. An overall definition would probably need to include such things as farm size, landownership and other agrarian questions.

The role of the farmer is changing, to include the general stewardship of rural areas and overall management of land, as well as producing food. If financial support for farming is continued it should be linked to this wider concept of stewardship, rather than simply to the maximization of food output. This would have several implications. One of them might be higher prices for food; differential taxes might be another. These should be taken into account.

A pragmatic shift of funding and support is needed for research and training, away from industrialized farming and towards organic farming. It is critical that in this process, the research, development and training does not all fall into the hands of academic bodies who have been working on intensive farming and who are now starting to change over. Rather a proportion should go to on-farm groups and local farmer co-operatives.

The role of genetic engineering, particularly gene-splicing, is a matter of concern. Participants opposed patent rights on living things and felt there was a need for an investigation into the role of transnational companies in genetic engineering.

Food security
Environmental degradation threatens food security in the future, but for the present there is enough food in the world; people starve because the food is not in the right place. Population growth cannot be ignored in food security matters. As it is linked with poverty, we have to tackle poverty if we want to do something about the growth of population.

Hunger in the South is connected to over-consumption

149

in the North, and to over-consumption by Third World élites, who are often supported by the policies of their own governments. The policies that the International Monetary Fund persuade developing countries to pursue on food and agriculture are often wrong. Governments would do better not to accept such advice.

It is not good for a country to depend too heavily on the export of crops. Countries should maintain higher levels of domestically produced food. While people are starving in Sudan, the country exports a considerable amount of cotton. In the interests of greater food security, a better balance is needed between export and food crops.

Organic, environmentally-friendly agriculture is a goal towards the imperative of sustainability, but, for the majority of farmers, time will be needed for a transition towards it. Participants also believed that the patenting of life will weaken food security in the South.

Notes

Foreword

1. *World Development Report*, 1990, World Bank, Washington, p. 65.
2. Frances Moore Lappe and Joseph Collins, *Food First*, Souvenir Press, London, 1980, p. 147.
3. H. Ruthenberg, *Farming Systems in the Tropics*, Clarendon Press, Oxford, 1980, p. 424.

Zimbabwe

1. The World Commission on Environment and Development report, *Our Common Future*, OUP, 1987.
2. Central Statistics Office (Zimbabwe) 1987, *Statistical Yearbook*, Harare, 1987.
3. 'Zimbabwe Agriculture Sector Review', IBRD, Washington, 1990.
4. 'Evaluation of EC Co-funded Resettlement Schemes in Zimbabwe', Final Report, GFA, Hamburg, 1988.
5. 'Land Resettlement in Zimbabwe: a preliminary evaluation', ODA Evaluation Report EV 434, Cusworth and Walker, ODA, London, September 1988.
6. Lionel Cliffe, 'Policy options for agrarian reform in Zimbabwe, a technical appraisal', FAO, 1986.

References

Berry, R. Albert and William R. Cline (1976), *Farm Size, Productivity and Technical Change in Developing Countries*, draft later summarized in *Land Reform in Latin America: Bolivia, Chile, Mexico, Peru and Venezuela*, Schlomo Eckstein *et al*, World Bank Staff Working Paper No. 275, Washington DC, April 1978.

Blaikie, Piers (1984), *The Political Economy of Soil Erosion in Developing Countries*.

Blake, Francis (1988), *Handbook of Organic Farming*, Crow Press.

Brandt, Willy (Chairman) (1980), *North and South: A Programme for Survival, The Report of the Independent Committee on International Development Issues*, Pan.

Brown, Lester R. (1976). *World Population Trends: Signs of Hope, Signs of Stress*, Worldwatch Paper No. 8, Worldwatch Institute, Washington DC.

Brown, Lester R. and Erik Eckhom (1974), *By Bread Alone*, Praeger Publishing.

Brown, Lester R. and Edward C. Wolf (1984), *Soil Erosion: Quiet Crisis in the World Economy*, Worldwatch Paper No. 60, Worldwatch Institute, Washington DC.

Bull, David (1982), *A Growing Problem: Pesticides and the Third World Poor*, Oxfam, Oxford.

Carroll (1969), Evidence to the US government given in *Proceedings of the Subcommittee on National Security Policy and Scientific Developments of the Committee on Foreign Affairs*, Ninety-first Congress, 5 December 1969.

CIIR (Catholic Institute for International Relations) (1987), *European Companies in the Philippines*, CIIR, London.

Cohen, John (1978), *Land Tenure and Rural Development in Africa*, Harvard Institute for International Development, Cambridge, Massachusetts.

Conway, Gordon R. and Edward B. Barbier (1990), *After the Green Revolution: Sustainable Agriculture for Development*, Earthscan, London.

Conway, Gordon R. and Jules N. Pretty (1991), *Unwelcome Harvest: Agriculture and Pollution*, Earthscan, London.

Dalby, John (1990), 'Have standards, will travel', *Living Earth*, July 1990, The Soil Association, Bristol.

Djigma, A., D. Lairon, E. Nikiema and P. Ott (1990), *Agricultural*

152

Alternatives and Nutritional Self-sufficiency, Proceedings of the IFOAM Seventh International Scientific Conference, Ougadougou, 2–5 January 1989, Ministry of Agriculture and Animal Husbandry of Burkina Faso and the International Federation of Organic Agriculture Movements.

Dinham, Barbara and Colin Hines (1983), *Agribusiness in Africa*, Earth Resources Research, London.

Dudley, Nigel (1987), *This Poisoned Earth*, Piatkus Press, London.

—— (1989), *Nitrates*, Green Print, London.

Eckholm, Erik (1979), *The Dispossessed of the Earth: Land Reform and Sustainable Development*, Worldwatch Paper No. 30, Worldwatch Institute, Washington DC.

—— and Lester R. Brown (1977), *Spreading Deserts: The Hand of Man*, Worldwatch Paper No. 13, Worldwatch Institute, Washington DC.

EFRC (Elm Farm Research Centre) (forthcoming), *Impact of Different Farming Systems on Populations of Earthworms*, EFRC, Berkshire.

Erlich, Paul (1968), *The Population Bomb*, Ballantine Books.

—— and Anne Erlich (1990), *The Population Explosion*, Arrow Books, London.

FAO (Food and Agricultural Organisation) (1989), *Land Reform*, FAO, Rome.

Farmers' World Network (1991), *Land and the Family Farm: A World View of Land Ownership and its Importance for the Family Farm*, Proceedings of a seminar held in Skelmersdale, 24 January 1991.

Farmers Assistance Board (1982), *Profits from Poison*, FAB, Manilla, Philippines.

Floquet, Anne (1991), Farmers' Assessment of Techniques, *ILEA Newsletter* 7 (1 and 2), Information Centre for Low External Input and Sustainable Agriculture, Leusch, The Netherlands.

Food Matters Worldwide (1990), Issue 10. Farmers' World Network/ Farmers Link, NEAD, Norwich.

George, Susan (1976), *How the Other Half Dies*, Penguin Books, Middlesex.

—— (1989), *A Fate Worse Than Debt*, Penguin Books, Middlesex.

Grainger, Alan (1990), *The Threatening Desert: Controlling Desertification*, Earthscan in association with the United Nations Environment Programme, London and Nairobi.

Harrar (1969) evidence to the US government given in *Proceedings of the Subcommittee on National Security Policy and Scientific Developments of the Committee on Foreign Affairs*, Ninety-first Congress, 5 December 1969.

Harris, Nigel (1986), *The End of the Third World*, Penguin Books, Middlesex.

Hayter, Teresa and Catherine Watson (1985), *Aid: Rhetoric and Reality*, Pluto Press, 1985.
Hines, Colin (1991), *the Four Horsemen of the Free Trade Apocalypse*, Earth Resources Research, London.
Holdcraft, Lane E. (1978); *The Rise and Fall of Community Development in Developing Countries, 1950–65: A Critical Analysis and an Annotated Bibliography*, Department of Agricultural Economics, Michigan State University.
Holmberg, Johan, Stephen Bass and Lloyd Timberlake (1991), *Defending the Future: A Guide to Sustainable Development*, International Institute for Environment and Development, London.
Hurst, Peter, Alastair Hay and Nigel Dudley (1991), *The Pesticides Handbook*, Journeyman, London.
INSAN (Institute for Sustainable Agriculture Nepal), *Annual Report 1989*, Kathmandu.
International Movements for Sustainable Agriculture (1990), *From Global Crisis Towards Ecological Agriculture*, Declaration of the International Movement for Ecological Agriculture, Penang.
IRRI (International Rice Research Institute), *'Azolla' helps organic farmer earn more*, Press Release IRRI, Manilla, Philippines.
Jackson, Ben (1990), *Poverty and the Planet: A Question of Survival*, Penguin Books in association with the World Development Movement, Middlesex and London.
Jodha, N. S. (1990), *Rural Common Property Resources: A Growing Crisis*, Earthscan, London.
Lampkin, Nicolas (1990), *Organic Farming*, Farming Press, Ipswich.
Lampkin, Nicolas and Peter Midmore (1989), in *The Potential for Organic Agriculture in European Extensification Systems*, Elm Farm Research Centre, Berkshire.
Lappe, Frances Moore and Joseph Collins (1976), *Food First*, Pan Books, London.
Lever, Harold and Christopher Hulme (1984), *Debt or Danger*, Penguin Books, Middlesex.
Living Earth (1989), magazine of the Soil Association, Bristol.
McRobie, George (ed.) (1990), *Tools for Organic Agriculture*, Intermediate Technology Publications, London.
Manor House Agricultural Centre Newsletter, Vol. 1, No. 1, November 1988.
Meadows, Donella L. *et al.* (1972), *The Limits to Growth, A Report for the Club of Rome*, Pan Books, London.
Mendosa, T. C. (1990), 'Development of organic farming practices for sugarcane-based farms', in Djigma *et al. Agricultural Alternatives*.
Myers, Dorothy (1990), *Implementing the Code*, The Pesticides Trust for PAN International, London.

Myers, Norman (1984), *The Primary Source*, W. H. Norton, Washington DC.

National Academy of Sciences (1988), *Alternative Agriculture*, NAS, Washington DC.

Newland, Kathleen (1979), *International Migration: The Search for Work*, Worldwatch Paper No. 33, Worldwatch Institute, Washington DC.

Ouedraogo, Abdoulaye (1990), 'Zai: traditional technique as a source of Sahelian soil productivity increase', in Djigma *et al. Agricultural Alternatives*.

Rist, Stephan (1991), 'Revitalising indigenous knowledge', *ILEIA Newsletter* 7(3), Information Centre for Low External Input and Sustainable Agriculture, Leusch, The Netherlands.

Roeleveld, Lex (1987), 'Xiao Hei Lei's farm', *ILEIA Newsletter* 3(4), Information Centre on Low External Input and Sustainable Agriculture, Leusch, The Netherlands.

Seagar, Joni (ed.) (1990), *The State of the Earth: An Atlas of Environmental Concern*, a Pluto Project for Unwin Paperbacks, London.

Stewart, V. I., J. Scullion, R. O. Salih, and K. H. Al-Bakri (1988), 'Earthworms and structure rehabilitation in subsoils and in top-soils affected by opencast mining for coal', *Biological Agriculture and Horticulture* 5 pp. 325–38.

ven den Bosch, Robert (1978), *The Pesticide Conspiracy*, Doubleday and Company, New York.

Vine, Ann and David Bateman (1981), *Organic Farming Systems in England and Wales*, Department of Agricultural Economics, University College of Wales, Aberystwyth.

Vogtman, Hartmut (1991), *Organic Agriculture and World Hunger, First Lady Eve Balfour Memorial Lecture*, The Soil Association, Bristol.

Walinsky, Louis J. (ed.) (1977), *The Selected Papers of Wolf Ladejinsky*, Oxford University Press, New York.

Watterson, Andrew (1987), *Pesticide Users Health and Safety Handbook*, Gower Press.

Weir, David and Mark Schapiro (1981), *Circle of Poison*, Centre for Investigative Reporting, San Francisco.

World Commission on Environment and Development (1987), *Our Common Future*, Oxford University Press, Oxford.